D0179523

Environmental Anthropology

Environmental Anthropology

From Pigs to Policies

Patricia K. Townsend
State University of New York, Buffalo

Prospect Heights, Illinois

For information about this book, write or call:

Waveland Press, Inc.
P.O. Box 400
Prospect Heights, Illinois 60070
(847) 634-0081
www.waveland.com

Photo Credits
Cover Palm tree and sunset, Tenerife, Canary Islands. *Chapter 1* The author in Yapatawi village, Papua New Guinea, 1967. Photo by William H. Townsend. *Chapter 2* Inuit fisherman. Photo by Ann McElroy. *Chapter 3* Culina man making arrows, Papua New Guinea, 1967. Photo by William H. Townsend. *Chapter 4* Saniyo widow, Papua New Guinea, 1967. Photo by William H. Townsend. *Chapter 5* Culina hunter and peccary, Peru, 1972. Photo by William H. Townsend. *Chapter 6* Andean peasants with llamas, Peru. SuperStock, Inc. *Chapter 7* Ok Tedi forest dieback. Photo by Stuart Kirsch. *Chapter 8* Inuit child. Photo by Ann McElroy. *Chapter 9* Macau, China. SuperStock, Inc. *Chapter 10* Sago palm forest. Photo by William H. Townsend. *Chapter 11* Arctic seal hunting. Photo by Ann McElroy. *Chapter 12* Alison Townsend with Culina child, 1972. Photo by William H. Townsend.

ISBN 1-57766-126-5

Printed in the United States of America

7 6 5 4 3 2 1

Contents

Preface

My interest in the subject matter of this book goes back to studying grade school geography with Miss Schuneman at one of the last one-room schools in rural southwestern Michigan—daydreaming by the creek in our back pasture about going to the faraway places I had studied. More directly, it dates from an excellent undergraduate education at the University of Michigan where the influence of Marshall Sahlins and Kenneth Pike converged in a few short semesters to set my intellectual course, even though both of my first anthropology teachers soon moved off in quite different directions than the one I have taken. The first six chapters of the book reflect that personal and disciplinary history, while the final chapters reflect my growing conviction that, of the various scholarly disciplines, anthropology has both an appropriate degree of humility and a broad enough vision to address the environmental mess that we humans have made. In writing this book I benefited from the encouraging words and critical comments of Tom Curtin, who suggested it, Bill Townsend, who supported it, Amy Grasmick, who repaired it, and also Eugene Hunn, Stuart Kirsch, Ann McElroy, Pete Vayda, Jim Weil, and two anonymous readers who are probably still not entirely satisfied.

Pat Townsend
Buffalo, New York
May 8, 2000

Chapter One

Introduction

When I went to live with the Saniyo-Hiyowe of New Guinea as a 24-year-old graduate student in anthropology, in many ways I had to begin as a baby. There were no dictionaries or grammars of their language. I could not study it in advance but only by bumbling along, pointing to a house or a pig and imitating the response, *"wesi"* or *"fei."* I had to hope that what I was learning was really the word "house" or "pig" and not "finger" or "what the heck does she want?"

The villagers had to teach me manners. It is rude to point with your finger; instead you gesture with your lower lip. You are not supposed to watch someone eat. You can pet the pigs, but you must kick the dogs, to discourage them from hanging around trying to grab food. Because I thought I was a reasonably good hiker, it was most humiliating to have mere children or old ladies try to give me a hand as I felt my way across bridges consisting of a single skinny log across deep mud holes. All of this to learn before I could even begin to record their knowledge of the tropical forest environment.

Eventually, in nearly two years of fieldwork, my husband, Bill, and I did collect a lot of plants for the national herbarium and created a glossary of plant and animal terms as part of my dictionary of the language. Bill, who is a civil engineer, mapped the immediate area and documented the use of stone tools to cut trees, discovering that it takes a man skilled in using both a stone tool and a steel axe about four times longer to cut a tree with the former as it does with the latter.

The most important part of the biotic environment for my study had to be the sago palm, *Metroxylon sagu*. It was the carbohydrate staple of the diet. Sago starch was a monotonous 85 percent of the food eaten. Many other foods were eaten in small amounts as merely a garnish or snack. These included fish, meat from wild and domesticated pigs, small game, insects, leafy greens, bananas, root crops such as cassava and taro, fruits, and nuts.

Preparing sago starch was the main work activity of the women. They cut, scraped, pounded, and washed the pith of the palm to extract the starch, carrying it home in big pinkish-white blocks wrapped in leaves. Most of the sago palms were of seedless varieties that must have at one time, perhaps centuries earlier, been planted from cuttings. These grew near the edge of the swamp forest, naturally regenerating the stand by sending up suckers at the base of each palm. Later, when I worked with three geographers on a book about sago, we subtitled it "a tropical starch from marginal lands" to stress how exquisitely adapted a crop it was for wetlands like these (Ruddle et al. 1978).

In the depths of the swamp there grew "wild" sago palms with seeds. They could be harvested for starch or to grow delicious fat grubs—beetle larvae. For all practical purposes, the Saniyo were foragers rather than farmers, seldom needing to plant anything other than bananas and tobacco.

Besides the role of sago in subsistence, another major issue that interested me in the field was population. Infant and early childhood mortality were very high, with just over half of the children surviving the first few years of life. The shortage of suitable weaning foods increased the children's vulnerability to infectious disease, though we rarely saw severe malnutrition. Malaria and respiratory diseases such as pneumonia, bronchitis, and tuberculosis were common causes of illness and death. The population was barely replacing itself with this high death rate. Fertility was not very high either. An average woman who lived past the age of 50 gave birth to 5 children, but many women died young, before the end of their potential childbearing years.

High mortality is not something that the people ever took for granted—they sought an explanation for every serious illness or death. The death of a child or an old person is attributed to one of several spirits. Some of these are the spirits of the recently dead or more remote ancestors. Some are spirits associated with features of the landscape: a large ironwood tree, a certain vine, or a whirlpool. Such spirit beliefs help to underwrite a certain attitude of respect for the environment. For example, my neighbors cautioned me not to laugh or speak loudly during an earth tremor or it might become a big earthquake.

If an adult dies suddenly in the prime of life, people claim that a witch is responsible. The local concept of a witch is someone who has eaten human flesh and acquired a taste for it. Witches consume other things that normal people wouldn't eat—earthworms, millipedes, and other repulsive creatures—and eat them through the nose and other orifices. Witches violate food taboos such as the severe ones imposed on widows and widowers. At first, after the death of a spouse, the bereaved can eat hardly anything, only a few bitter wild yams. Gradually vegetable foods are allowed but most meat remains forbidden for years. Some widows in our village could only eat nuts with their sagu. The Saniyo characterize a witch by his or her illegitimate, greedy consumption in violation of food taboos.

The high mortality experienced by the people we lived with, as in other parts of Papua New Guinea out of the reach of regular health services, reflected, in part, the effects of epidemic diseases such as influenza and whooping cough that swept through the area every few years. The spread of new epidemic diseases was one of the first ways that the Saniyo experienced the expansion of the global economic system.

In addition to epidemic disease, some plants that came from the Americas, such as tobacco and a seed used for its red pigment (*Bixa orellana*), also came into the area ahead of direct contact with Europeans. Many new food plants arrived (squash, papayas, pineapples, lemons), as well as new animals (cats, chickens). An exotic African species of fish, tilapia, that was bigger than any of the native species was introduced into the Sepik River and worked its way up the tributaries.

Arriving five years after the first colonial Australian patrol, we were there to witness a time of rapid expansion of the world economic system into the area. Steel axes and knives were replacing stone, shell, and bamboo. Used cotton and polyester clothing was being traded into the area, replacing scratchy grass and leaf skirts. A few men went out to coastal plantations to sell their labor. Oil and mineral exploration companies began exploration. They found gold and copper prospects nearby, but low mineral prices in the 1980s and 1990s made it unprofitable to consider mining them yet. Just as soon as prices rise significantly, the nearby Frieda Mine is likely to go ahead, bringing more dramatic change to the Saniyo.

THE FIELD OF ANTHROPOLOGY

The research that I did in New Guinea is called ethnographic fieldwork and the written results are called *ethnography,* quite literally "writing about people," or, the scientific description of a culture. When I introduce myself as an anthropologist, people invariably say, "Oh, how interesting, you study old bones and stones," and I mumble, "No, I'm a *cultural* anthropologist. I study *living* people." But they are not far off the mark in naming two of the main subfields of anthropology. The "bones" as well as flesh, blood, and genes refer to *biological anthropology,* also called physical anthropology or human biology. The "stones" refer to *prehistoric archaeology,* the study of the material remains of past cultures. The remaining, fourth traditional subfield is *anthropological linguistics.* A strong training in linguistics was especially important for me because I was going to a place that was a big blank on the language map of the world. A fieldworker who planned to go to Japan, for example, would find already-existing grammars, dictionaries, and even language courses allowing language learning before going to the field, which was not possible for me.

To the traditional four subfields of anthropology, there is sometimes added a fifth, *applied anthropology.* Others simply regard applied anthropology as a dimension of all the subfields, that is, any use of anthropology to solve practical problems. This is what I have done for much of my career, for example, studying health service delivery in Papua New Guinea from 1980 to 1984. Later I directed a church-related agency that resettled Kurdish, Sudanese, Russian, and other refugees in Buffalo, New York. Most recently I have done work on the role of churches and interfaith coalitions in the communities surrounding Superfund sites and their role in identifying hazards, communicating risk, and remediating these hazardous waste sites. The Superfund is a federal program of the U.S. Environmental Protection Agency for cleaning up the nation's most seriously chemically contaminated sites.

Threaded through all of the subfields of anthropology is environmental anthropology, the topic of this book. My own training is largely in cultural anthropology, so the fieldwork I did in Papua New

Guinea and the current work I am doing in Superfund communities illustrates how a cultural anthropologist goes about studying environmental issues using the methods of ethnography—interviewing and observing people's behavior. An archaeologist uses quite different methods, digging up the evidence of human impact on ancient environments. Because archaeological evidence can cover very long periods of time, archaeologists can also tell us how very gradual changes in climate led to changes in culture. One of the most imaginative uses of archaeological methods is the University of Arizona's Garbage Project (Rathje and Murphy, 1992). Archaeology students sorted bags of household garbage. Later they also excavated landfills. Initially intended merely as a way to learn archaeological techniques, the project gave unusual insight into our culture's patterns of consumption, waste, and recycling. For one thing, they discovered by sorting the hazardous wastes in household garbage that people throw out a lot more household cleaners and pesticides than they suspect or admit to. In subsequent chapters of this book there will be examples of environmental anthropology that draw on all the subfields of anthropology.

ORGANIZATION OF THIS BOOK

This book will follow a somewhat historical path through the first several chapters although it is not intended to give a full-fledged history of environmental anthropology. Chapter 2 picks up the history of the field in the late 1940s and early 1950s, when Julian Steward introduced the concept of *cultural ecology* into anthropology. Chapter 3 moves on to the early 1960s, when much of the excitement in environmental anthropology surrounded *ethnobiology*, a new approach to fieldwork that drew mostly on linguistics to study the traditional knowledge that people had about plants, animals, and other aspects of the environment. Chapters 4, 5, and 6 are concerned with work that started in the 1960s and 1970s and brought concepts and methods into anthropology from outside the discipline, primarily from general ecology as practiced by biologists. Chapter 4 highlights the work of Roy Rappaport in New Guinea and chapter 5 the work of several anthropologists in the Amazon region. All this research in small tropical forest communities is balanced in chapter 6 by looking at the ways that ecological anthropologists approach larger agricultural populations in more complex societies. This chapter refers to classic studies by well-known anthropologists such as Fredrik Barth, Clifford Geertz, and the late Robert Netting and Eric Wolf. Throughout this roughly historical overview, in chapters 2–6 I emphasize that many of the ideas and approaches developed earlier continue to be useful in the present, contributing to understanding the ways that present-day local populations of human beings adapt to their physical and biological environments.

The later chapters of this book explore the interaction between the local and the global that preoccupies environmental anthropologists today. Chapter 7 tells about the social and environmental impact of an open-pit copper and gold mine. While this one happens to be in Papua New Guinea, the multinational corporation that owns it also operates mines in Australia, Canada, and Chile, making these issues truly global. Chapter 8 explores the concepts of risk and hazard as they apply to technologies as different as fishing nets and nuclear power plants. Chapter 9 considers world population from the angle of vision provided by anthropology, a discipline that is strongly influenced by its fieldwork in local populations, particularly those where disease and poverty are prevalent. Chapter 10 looks at the global loss of biodiversity and its implications for human health.

The final two chapters consider how, as professionals and as individuals, we can move toward personal engagement with environmental problems. Chapter 11 describes the work of applied anthropologists on practical environmental problems such as Indonesian forest fires. Chapter 12, the concluding chapter, talks about personal lifestyles and their implications for the global environment.

Chapter Two

Julian Steward's Cultural Ecology

No species thrives in such a wide range of environments as Homo sapiens. Spreading out, often at the expense of other animals and plants, we humans have come to occupy lands with a wide range of latitude and altitude, with varying physical characteristics and rainfall. From fossil remains several million years old, physical anthropologists infer that we emerged in a subtropical African savanna, a landscape of mixed trees and grassland. From there we have gone on to inhabit deserts, tropical and temperate forests, high altitudes above the tree line, and most recently the most extreme Arctic conditions. In our century, we have even made short visits to outer space and undersea environments.

The adaptations that made it possible for humans to spread into so many different environments by and large have not been genetic ones. Homo sapiens remain a single, interbreeding species with relatively minor differences among us to show for our past experience with environmental extremes. Pale, lightly pigmented skin, for example, is an adaptation that allowed people to avoid vitamin D deficiencies (the crooked bones of rickets) in northern environments lacking adequate sunshine. The cost of this adaptation was an increased susceptibility to skin cancer after light-skinned people again migrated to sunny climates like Australia and California. Similarly, the sickle cell trait and certain other variants of hemoglobin are genetic adaptations that offer some protection against malaria. They offer survival advantage only in Africa and other places where the parasites that cause malaria and the mosquitoes that transmit malaria are part of the environment.

The adaptations that allow Homo sapiens to thrive in many different environments are largely behavioral, not genetic. We build houses suitable to the climate, put on or take off clothing, make and use tools that are efficient for capturing whatever other species are available and desirable for food. Most of these behavioral adaptations are socially learned; that is, they are cultural. And culture is traditionally the main subject matter of anthropology. In this adaptive perspective, cultures are seen as intimately related to the physical and biological environments in which they occur. A culture is a way of life, a tool kit for survival in a particular place on the planet.

A SHORT HISTORY OF ENVIRONMENTAL ANTHROPOLOGY

Environmental perspectives go back a long way in the history of anthropology, but they have had their ups and downs. In some eras, anthropologists have been very interested in the environment; at other times they tended to contemplate culture as a thing-in-itself, like a beautiful vase on display or a written text to be translated or interpreted without much reference to its environment. In this short book, we will not look back farther than the middle of the twentieth century, when Julian Steward introduced the idea of cultural ecology. Steward's cultural ecology gathered adherents and then was transformed into ecological anthropology. This transformation came in part from adopting the concept of the ecosystem from biology.

Ecological anthropology got a big boost from developments in the wider society of the United States and Europe in the 1970s. After waiting in long lines at gasoline pumps, many people were worried that population growth and economic development would outrun the supply of petroleum and other nonrenewable resources. Also in the late 1970s the publicity surrounding the health effects of industrial wastes at sites such as Love Canal in Niagara Falls, New York, made it clear that the problems of pesticides and other hazardous chemicals affected people directly. Hazardous wastes could not simply be buried and forgotten. The concerns about the dangers of pesticides that Rachel Carson had publicized in her book *Silent Spring* (Carson 1962) were not limited to birds. With public interest in environmental issues at a peak, this period of the 1970s was also a high point of research in ecological anthropology. This work culminated in a series of textbooks that summed up the research and concepts of ecological anthropology to date (Bennett 1975; Ellen 1982; Hardesty 1977; Little and Morren 1976; Moran 1979).

Most of these textbooks did not go into second editions in the 1980s because the next decade was a time when anthropology was not paying much attention to environmental questions. Partly this reflected a loss of public interest. These were the years of the Ronald Reagan presidency, when much of the environmental legislation passed in the 1970s was weakened by deregulation and lack of enforcement. Partly the development was internal to anthropology; other questions were more popular with researchers. Even those anthropologists who continued to work on environmental matters took some time to question and rethink their concepts and methods.

By the 1990s, environmental anthropology was again at the forefront. As always, developments in the academic disciplines paralleled those in the wider public. The United Nations Conference on Environment and Development held in Rio de Janeiro, Brazil, in 1992 drew public attention to issues such as deforestation and the loss of biological diversity. Another United Nations conference in Kyoto, Japan, in 1997 gave voice to concerns about global climate change and called for a 55 percent reduction in carbon dioxide emissions. In the American Anthropological Association, the major scholarly and professional organization for the discipline, a new section called Anthropology and Environment was organized. Several graduate departments, such as those at the University of Georgia and the University of Washington, established programs with concentrations in environmental anthropology. Most importantly, environmental anthropology was no longer just an academic discipline discussed by researchers, college professors, and students. The conversations now included many practitioners—applied environmental anthropologists working in government agencies, nongovernment organizations (NGOs), and businesses. Their research was designed to help formulate environmental policies and programs.

Anthropologists' involvement with environmental issues has become more differentiated, with a wide variety of methods, theories, and specialized research interests. Anthropologists using these newer approaches might identify themselves as evolutionary ecologists, historical ecologists, political ecologists, or ethnoecologists. In this text, I use the term *environmental anthropology* as an umbrella for all of these approaches.

Some anthropologists use the term *ecological anthropology* for what I have called *environmental anthropology*, but I will use the term *ecological anthropology* in a somewhat narrower sense. *Ecological anthropology* will refer to one particular type of research in environmental anthropology—to studies that describe a single ecosystem that involves a human population. Studies in ecological anthropology usually deal with a small population of only a few hundred people.

THE CULTURAL ECOLOGY OF JULIAN STEWARD

More than any other anthropologist, Julian Steward was responsible for the development of environmental anthropology, particularly in the years he spent at Columbia University in the late 1940s and early 1950s, where he influenced an important generation of anthropologists with his theory of cultural ecology and evolution. Steward had started his career at the University of California in the 1930s. At that time he made several trips throughout the Great Basin area of Nevada, Utah, and parts of neighboring states, visiting the Western Shoshoni and their Paiute and Ute neighbors.

By the time Steward did his survey, the indigenous societies of the Great Basin had been drastically affected by the intrusion of miners and ranchers. Grazing sheep and cattle had reduced the wild seeds that had been the mainstay of the traditional native diet. Large game had always been scarce in this area, though occasionally deer, mountain sheep, bison, or antelope could be killed. Rabbits and other small game were more significant. Seasonal migrations were determined by the plant life—greens in spring, ripening seeds in early summer, edible roots and berries in late summer, and pine nuts in the fall.

After their nineteenth-century wars with the settlers, the Indians left their old system of subsistence based on foraging on the land to work on ranches or in mines and towns. Because of this change, Steward could not do intensive fieldwork to observe ongoing subsistence activities, that is, the kind of fieldwork we will see in later chapters of this book. Steward patched together nineteenth-century descriptions and censuses, his own plant collections and environmental observations, and data he collected by interviewing Indians, such as their lists of the names and uses of plants.

During the pre-settler period, most of the subsistence activities of the Shoshonean Indians were carried out by one or two families moving about this dry, sparsely populated area to forage for plant foods and small game. Only in the winter were they free to camp with 20 or 30 families living close enough to visit each other.

Steward rejected the notion, prevalent in his time, that the culture of the Shoshoni could be explained only by tracing historical links to earlier cultures. While he agreed that their material culture—baskets, pots, and tools—was mostly derived from the Southwest, he considered that their economic and social organization was the result of using that technology to exploit a particular environment, the arid Great Basin with its unpredictable resources. The features of social and economic life that are most closely related to subsistence were, in Steward's term, part of the *cultural core* (Steward 1955).

In approaching a new culture, Steward said, it is not possible to know what is in the cultural core in advance, but by using the *method of cultural ecology* one can determine this. First, the anthropologists analyze the relationship of the technology used in production to the environment in which it is used. Then they relate other behavioral patterns to subsistence. For example, do people work alone or cooperatively? Finally they can ask how these behavioral patterns affect other aspects of the culture, such as kinship, warfare, or religion.

EVOLUTION IN CULTURAL ANTHROPOLOGY

Steward's method of cultural ecology was part of a more general move beginning in the 1940s to reintroduce the concept of cultural evolution to anthropology. Nineteenth-century anthropologists such as Lewis Henry Morgan and E. B. Tylor had advocated the view that all cultures evolved through a similar series of stages from simple to complex. American anthropology in the early twentieth century unfolded in reaction to this view. This reaction took the form of *cultural relativism*. Each culture was accepted on its own terms as a product of its unique history. Attempts to find general laws or causal explanations were regarded with suspicion. Steward disagreed, arguing that there were regularities to be discerned in the way that cultures change. It was the job of anthropology to determine these cause and effect sequences through empirical study, that is, scientific research based on observation and comparison. These sequences of change were not universal, hence Steward used the term *multilinear* evolution in contrast to the *unilinear* evolution of nineteenth-century anthropology.

Many of Steward's students and other anthropologists picked up the concept of multilinear evolution but generally they called it by other names. Elman Service and Marshall Sahlins (Sahlins, Service, and Harding 1960) distinguished two processes of cultural evolution—general and specific. What they called specific evolution was the equivalent of Steward's multilinear evolution and also came to be called "cultural adaptation." This idea that a culture adapts to its biophysical environment became a major theme in anthropology, a theme that anthropologists alternately attempt to demonstrate or disprove. Because existing cultures must be somewhat adapted or the people that practice them would have died out, some critics argued that adaptation is simply a truism: whatever exists is adaptive.

Other critics pointed out that it is not difficult to see how *mal*-adapted much of human culture is. In our own culture we wear earrings or nose rings in holes that become infected, wear high heels that throw our skeletons out of alignment, or smoke cigarettes that damage our heart and lungs. Many other cultural practices lead to disaster and death in "sick societies," as Robert Edgerton (1992) called them. Islanders hunted bird species to extinction, people spread infectious disease by cannibalism or sexual promiscuity, and others deprived children of needed protein by food taboos.

It does not seem to be particularly helpful to concentrate on the *outcome* of adaptation. Instead it is more helpful to concentrate on the *process* of adaptation, accepting that the outcome may not always be favorable (*mal*adaptation). Cultural ecologists who follow in Steward's path today look at the way societies respond to changes in their environment and in the cultural core.

CULTURAL ECOLOGY AND THE LAST NORTHERN COD

A contemporary example of cultural ecology comes from the Canadian fishing industry. The northern cod (*Gadus morhua*) became so scarce in the Atlantic waters off Newfoundland that the fisheries management officials had to declare a moratorium on fishing in 1992. When the cod-fishing industry collapsed, thousands of fishing crews and plant workers lost their jobs. Most social scientists tried to explain this collapse by reference to economic and political factors. One of the anthropologists who analyzed the catastrophe found that the more important thing to understand was the relationship of specific aspects of fishing technology (i.e., Steward's culture core) and environment (McGuire 1997).

Foreign factory freezer trawlers from many European countries began fishing intensively in the waters off Newfoundland in the 1960s. Regulation of their catch was ineffectual, and the catch declined steeply from a peak of 800,000 tons in 1968 to about one quarter of that in 1976. In 1977 Canada barred the foreign fleets, declaring exclusive control over its coastal waters in a 200-mile zone. The Fisheries Department took on regulation of the industry. They limited fishing by issuing fishing licenses and keeping the total catch at a level set by the best scientific knowledge available. Unfortunately, the method for assessing catches turned out to be flawed. By turning to the "hard" sciences, the fisheries agency ignored the local knowledge of inshore fishing crews. The crews had already picked up danger signals that the fish stock was in trouble in the 1980s, when they had to work harder to catch smaller fish than previously.

Changes in the technology of fishing accompanied the declining catch. Increasingly the Canadian crews used Japanese cod traps, roofed fish traps with smaller mesh that could be used in areas with rough bottoms. These traps captured the smaller, younger fish. Other new technology included echo sounders and advanced navigation equipment. All of these changes in technology made it possible to catch "the last northern cod." At the time of this writing, most scientists expected the fish to become extinct.

As the northern cod example shows, Steward's approach continues to be useful. Many anthropologists continue to identify themselves as cultural ecologists, using Steward's term, right up to the present day (Anderson 1996; Netting 1986). They keep both halves of that term, regarding both culture and ecology as of equal significance. Not surprisingly, however, cultural ecology branched almost immediately into one wing of anthropologists who put more emphasis on the "cultural" (chapter 3) and another wing who put more emphasis on the "ecology," arguing that culture was a term that had outlived its usefulness to anthropology (chapter 4). They identified populations, not cultures, as the adaptive units that respond to the environment and form the basic units of the ecosystem.

Chapter Three

Ethnoecology

Culture has long been the key concept in American anthropology, and it has had almost as many definitions as there are anthropologists, though these definitions share many common features. One recent definition is: "Culture is what one must know to act effectively in one's environment" (Hunn 1989), where the word *environment* has both natural and social components. The environment is seen from the point of view of an individual actor. The focus is on culture as a system of knowledge or a set of rules for behavior, that is, on the *cognitive* aspects of culture. A deliberate ambiguity in the quoted definition is the word *effectively*. Who is to judge what is effective? An insider (a native of the culture), an outsider (the scientist), or the impersonal processes of natural selection and evolution that weed out the unsuccessful? It may be that these will coincide most of the time, as individuals pursue cultural goals that also lead to survival of the biological species.

One whole branch of environmental anthropology has followed the path suggested by this definition of culture, developing increasingly sophisticated studies of cultural knowledge. Cognitive anthropologists developed their methods by studying relatively small and closed sets of terms such as the words used for naming colors or classifying kinship relations. Each of these sets of terms is called a *semantic domain*, that is, an area of meaning in a language. Some cognitive anthropologists studied features of the physical environment such as the 13 types of ice recognized in a Native Canadian language of the far north, where the kind of ice is important to making decisions about safe travel across a frozen lake or the likely success of a fishing trip. But the studies that are most numerous and relevant to us are studies of taxonomy, that is, the way that people name and classify plants and animals. This kind of study is called *ethnoecology*.

THE STUDY OF TRADITIONAL ENVIRONMENTAL KNOWLEDGE

Ethnoecology and ethnobiology are general terms that cover the field. More specifically, studies dealing with traditional knowledge of plants are called ethnobotany, of bugs, ethnoentomology, and so on. To accomplish an ethnoecological or ethnobiological study, researchers begin with the native language categories. They need to know something about linguistics. They also need to gather a good deal of information from both observation and from interviews and texts in order to understand what differentiates the plants or animals named. Usually ethnobotanists make plant collections for a herbarium to be certain about the botanical identification. Ethnozoologists sometimes collect specimens by trapping animals or purchasing skulls from hunters in order to continue study of the bones and teeth back in the lab to confirm identification of the species.

Ethnobiology is not just a matter of matching up scientific labels to the local language names but also of understanding what features of the plant or animal people attend to in classifying them as they do. These may be quite different from what scientists emphasize. From these categories, the ethnoscientist moves on to understanding the cultural rules for using them—recognizing what reeds and palm woods are useful for making arrows of specific types for hunting birds, mammals, or other game, for example. Ultimately the ethnoscientist seeks to understand how people make complex decisions such as where to clear a field and when to plant, weed, and harvest.

Speakers of the Sahaptin language along the Columbia River of Washington and Oregon use 26 recorded words for different kinds of roots, *xni-t*, "plants that are dug" (Hunn and Selam 1990). Most of the 26 different named types of roots correspond fairly closely to biological species as scientists classify them. But at the next step up in hierarchy of the Sahaptin classification the more general ethnobotanical category "roots" includes very different members of several plant families. Some of the roots belong to the family Umbelliferae, a large botanical family that includes familiar plants such as carrots, parsley, and celery. Others, including the Sahaptin staple food, the camas, belong to the Liliaceae, the family that includes onions and garlic as well as lilies. These "roots" are closely related, from a botanist's point of view, to other plants that Sahaptin speakers know to be inedible or even poisonous. Despite the botanical similarities, the Sahaptin do not include the inedible plants in their category xni-t.

Columbia River Indian elder James Selam and his extended family taught their rich cultural knowledge of the environment to anthropologist Eugene Hunn. Among the hundreds of plants and animals documented, the roots had been particularly important in their ancestors' traditional diet, along with salmon and venison. This knowledge was valued and retained by some modern-day Sahaptins, although their lives and diets were much changed by the establishment of reservations and the building of hydroelectric dams along the Columbia. Hunn's work with James Selam on roots is a good illustration of ethnoecology. Other similar studies could be cited. The pioneer of the field was Harold Conklin (1954), who described the knowledge required by the Hanunóo of the Philippines to farm the tropical forests.

As ethnoecologists completed studies of numerous folk classification systems, it became possible to reach conclusions based on the comparison of more than one such system. This method can be used, for example, to address the question of differences between foragers and farmers in their knowledge of the environment, as it is reflected in their botanical terminology. The Guajá and the Ka'apor both live in the eastern part of Brazilian Amazonia. They inhabit the same botanical habitat and speak closely related languages. Yet the Ka'apor, who are horticulturalists, have many more botanical terms than the Guajá, who are foragers. This difference does not only reflect the fact that the Ka'apor have more terms for cultivated plants; they also have many more differentiated terms for kinds of wild plants, about five times as many. How can this be? As tropical forest farmers the Ka'apor are familiar with the kinds of trees that they cut to make their gardens. Their gardens create light gaps in the forest that make a highly biologically diverse habitat. When their crops fail, they are prepared to forage for wild plants, too. For all of these reasons they have developed and preserved an even richer botanical vocabulary than the Guajá (Balée 1999).

An important achievement of ethnoecological studies is that they bring recognition to the traditional environmental knowledge of indigenous peoples, who are often ethnic minorities held in contempt by racists among the majority populations of their country. Their subsistence systems are often criticized, too, by outsiders who see them as backward and who covet their land for raising cash crops or more intensive farming. Following ethnoecological studies, it is clear that traditional environmental knowledge is a body of knowledge that is extensive, observationally grounded, and complementary to scientific knowledge.

Tragically, these systems of traditional environmental knowledge are threatened not only by the loss of lands but also by the loss of the languages that encode them. At the current rate of change, at least 3,000 of the world's 6,000 languages will become extinct within the next century. Already at least 20 percent of the world's languages are dying. They are spoken only by old people, and children are not learning them; instead they are learning national or trade languages (Krauss 1992).

NATURE AND CULTURE: A UNIVERSAL DICHOTOMY?

Underlying our discussions of folk taxonomies so far has been an unstated assumption that there is something called "nature" in contrast to "culture." Nature is "out there" and consists of the biotic and abiotic (living and nonliving) environment. Culture is "in here," in the human mind, but we can find out about it by asking the right questions and observing behavior. There is a hierarchical relationship between culture and nature: Culture has dominion over nature and imposes order on it.

The nature:culture distinction once was unquestioned. The French anthropologist Claude Lévi-Strauss saw it as a universal structure, hardwired in the human brain. From the binary contrast between nature and culture, Lévi-Strauss thought that other metaphorical contrasts followed logically, such as the dichotomy between wild and tame, raw and cooked, and female and male.

This assertion that women were associated with nature and men with culture led feminist anthropologists to question the assumption that the distinction between nature and culture was universal (MacCormack and Strathern 1980). MacCormack's work in Africa and Strathern's in New Guinea led them to suggest that the concept of "nature" was ethnocentric, that is, that Europeans and Americans simply were imposing their own ideas about nature on other people. Other cultures did not have the same idea of nature or the metaphors that were said to follow from it. Other cultures did not separate the cultural from the natural. Even if they did make that separation, they constructed the boundaries between them differently than we did. After the feminist anthropologists opened up this topic of nature and culture, anthropologists continued to debate it fervently.

In the 1990s it became common for anthropologists, under the influence of the intellectual fashions of the day, to say that nature is "socially constructed." By this they did not mean that there was no biophysical reality or that it was all in our (collective) heads. They did insist that the language we use about nature does not simply mirror an essential reality out there but brings it into being and shapes it into a social reality. Different cultures use different symbols or metaphors in talking about nature, for example, many foraging people describe the forest as a parent, giving them sustenance, while scientists in the electronic era, including ecological anthropologists, describe the human environment with analogies such as thermostats and computer programs. Talking about nature is not just a matter for linguists to analyze but also raises questions of power, of who has the authority to speak (Escobar 1998).

After a promising beginning in the 1960s, ethnoecology remained an inconspicuous interest of relatively few anthropologists for several years. It seemed to be an esoteric research method without practical application to environmental problems. Ethnoecologists themselves fostered this impression by emphasizing the academic rather than practical aspects of their work. They hoped to influence the way that all ethnographers did fieldwork by the example of their studies.

Within environmental anthropology, ethnoecology was overshadowed by studies that downplayed the "cultural" in "cultural ecology." These were studies of ecosystem (discussed in chapter 3) and evolutionary ecology (discussed in chapter 4). True, a few ethnoecologists were widely admired for their mastery of a specialized domain of natural history. Anthropologist Ralph Bulmer and his Kalam collaborator Saem Majnep gained respect for work on New Guinea birds, for example. So did Brent Berlin and his colleagues for work on Tzeltal plants in southern Mexico.

As we will see in chapter 8, interest in ethnoecology began to flourish again when tropical deforestation, habitat loss, and the loss of biodiversity became important public concerns. Meanwhile, ethnobiologists had started to make significant contributions as applied anthropologists. One ethnobiologist served as an expert witness in a federal court proceeding on behalf of Native Americans seeking to protect their treaty rights to salmon and steelhead fisheries (Hunn 1999). Others helped to show how traditional ecological knowledge of indigenous people in areas like the Amazon could be used to design programs of economic development. It was hoped that they might be more likely to succeed than past environmentally destructive projects such as mining and large hydroelectric dams (Posey, et al. 1984).

Chapter Four

Pigs for the Ancestors

As their pig herd increased in size, quarrels erupted among the Tsembaga Maring of the mountainous interior of Papua New Guinea. Pigs foraging in the tropical forest occasionally broke through a fence into a garden and uprooted sweet potatoes and yams. Women began to grumble about the extra work they had to do feeding the pigs, now that there were too many pigs to get along on just sweet potato peelings and leftovers: by June 1962, the herd numbered 162. There were almost as many pigs as there were people; the population of the Tsembaga local group was 200.

The Tsembaga reached consensus that it was time to hold a *kaiko*, a yearlong festival that would culminate in a huge feast at which most of the pigs would be sacrificed, leaving only 60 juvenile pigs by the end of 1963. The goal of the rituals was to increase the fertility and growth of people and their gardens and pigs. The pigs slaughtered in the rituals are dedicated to various spirits, particularly the spirits of deceased ancestors. Prior to the peace enforced by the Australian colonial government, the ritual cycle was integrated with the cycle of war and peace among Maring local groups.

During the year that the kaiko was held, Roy A. Rappaport and his wife, Ann, were resident with the Tsembaga. They observed the rituals as well as the routine of daily life. Rappaport was at the time a Ph.D. student at Columbia University. As part of his dissertation research project he weighed all the food eaten by the Tomegai, a Tsembaga clan consisting of the people eating at four hearths: 10 adults, 1 adolescent, and 5 children. He also measured their gardens and observed gardening practices. He found that the main starchy staples that they grew—taro, sweet potato, yam, and banana—comprised more than half of the Tsembaga diet. Meat provided only about 1 percent of the normal day-to-day food. Even so, the Tsembaga diet was less monotonous than others reported from Highland New

Guinea. Variety was added by the abundance of greens, pandanus palm fruits, and sugarcane.

Though meat played little part in the day-to-day diet of the Tsembaga Maring, this changed radically at the culmination of the kaiko, when enough pigs were killed to provide every Tsembaga man, woman, and child with about 12 pounds of pork. The pork was eaten over a period of five days. In addition, 2,000 to 3,000 other people in neighboring groups received an average of two to three pounds per person. The total was nearly four *tons* of edible meat.

A NEW GUINEA ECOSYSTEM

Like the cultural ecologists who preceded him, Rappaport regarded subsistence as central to his research, though he avoided the concepts of culture and cultural core, preferring to speak of behavior, in order to facilitate comparison with the behavior of other animals. He went farther than the cultural ecologists had done in quantifying food and work. These measurements gave him the data needed to describe the flow of energy and materials through the ecosystem. Because he wanted to unify ecological anthropology with general ecology, he intended to use the concept of ecosystem in the same way that a biological ecologist would use it to describe an ecosystem such as a pond or a forest. The ecosystem concept was derived from the work of ecologists such as Eugene Odum. Odum's 1953 textbook *Fundamentals of Ecology* first promoted the concept widely, though it had been used earlier.

Rappaport considered the Tsembaga territory to be an ecosystem consisting of a population of human beings, local populations of pigs and other animal and plant species, and the nonliving substances such as soil and water on which they all depended. All of these components were linked into a system by the mutual exchange of matter, energy, and information—from the plants that produced food through photosynthesis to the bacteria that decomposed leftover sweet potato peelings.

Like the biological ecologists of his day, Rappaport employed a systems theory approach, treating the Tsembaga ecosystem as a more or less isolated, closed feedback system in equilibrium with its surroundings. Such a system is self-regulating and—using today's more fashionable term—sustainable. It was not assumed that the system would be changeless; indeed it had to change, adapt, or evolve simply to survive in a changing world. The use of systems theory was inspired by the enthusiasm for computer modeling, one of the cutting-edge developments of the time, though Rappaport's book did not itself contain a computer model. Rappaport's work stimulated a huge amount of discussion, re-analysis, and fieldwork by other anthropologists.

CRITIQUES OF RAPPAPORT'S WORK

Nutritionists pointed out that the kaiko, as described by Rappaport, was a very wasteful way to distribute the scarce animal protein available to the Tsembaga. Physiologically, they would have made better use of the protein by slaughtering fewer pigs on more frequent occasions. Of course, Rappaport had not claimed that this was the most efficient use of the meat; he only claimed that the timing of the ritual in coordination with the cycle of war and peace assured that high-quality protein would be available to injured warriors when needed. Nonetheless, this is probably one of the weaker links in his argument, as he himself admitted (Rappaport 1984, pp. 473, 478). The core of his argument was not the nutritional advantage but the role that ritual played in regulating the relationships between local groups.

Another set of criticisms that responded to Rappaport's book had to do with setting of boundaries around an ecosystem, in both space and time. How could Rappaport limit his study to the Tsembaga when they fought and traded with other local groups? This problem is one that biologists face as well. What are the boundaries of a forest or a stream? The scientist sets these boundaries somewhat arbitrarily, but only after considering whether the exchanges within the system are greater than those outside. Some ecological anthropologists after Rappaport despaired of defining higher-level units such as ecosystems and chose to focus instead on the individual organism interacting with its environment and on the process of natural selection operating on individuals. These anthropologists made their interdisciplinary connections more to evolutionary biology than to ecology. Rather than studying whole ecosystems and assuming that they are self-regulating, they chose to study the payoff of different behavioral strategies adopted by individuals. This approach was most easily applied to studying the strategies adopted by hunters (Alvard 1995; Smith 1991).

Setting limits of time is just as tricky as setting limits of space. Rappaport's study was based on data from a field study of a little more than a year and on people's memories of a generation or two. This is not a great deal of time depth. Tsembaga Maring ritual is assumed to work as a homeostat; that is, it returns the system to stability when it threatens to get out of balance. (For a comparison, think of a thermostat that switches a furnace off and on in order to keep the room temperature within comfortable limits.) But how did this system of ritual regulation come to be? How can this analysis allow for change and evolution? In the years just after Rappaport's study, cultural anthropology in general moved into a phase of great interest in history. Ecological anthropologists also began to accommodate longer, historical time frames and environmental transformations (Crumley 1993).

Rappaport wanted to link ecological anthropology to general ecology. In doing so, he linked it to the ecology of his day, but ecology as a science was changing, too. Ecology began to move away from the emphasis on equilibrium that Rappaport had picked up. The titles of papers and books in the field now use words like "chaos," "discord," and "surprise." Many scientific disciplines in addition to ecology have recently moved toward nonequilibrium models. When geomorphologists discuss the form of rivers, for example, they now emphasize initial conditions (bedrock) and cataclysmic events (super-floods) as determining the landscape. This is a change from the equilibrium approach that had dominated the field for several decades. In anthropology, the parallel to this approach is to look to history and unique events rather than the everyday processes that are ongoing when the anthropologist happens to be in the field.

One of Rappaport's lasting contributions was his rejection of the prevailing view within the social sciences that the only function of religion was to bind the community through socially shared symbols.

This century-old view was proposed by the French sociologist Émile Durkheim. Rappaport's study of the kaiko showed that rituals also had measurable material effects in ecosystems. They regulated such undeniably solid things as pig populations, the frequency of war, and the ratio of people to land. This utilitarian view of religion was disparaged by some anthropologists, but it also led to a new respect for the ability of traditional systems of knowledge to regulate environmental relationships. Rappaport also disagreed with the notion that religion was no more than an illusion that was bound to disappear as reason and science advanced. Religion has clearly persisted in today's societies, except for small numbers of outspoken skeptics.

After his initial fieldwork with the Tsembaga Maring, Rappaport continued to use the concept of ecosystem in his work, but he came to be less interested in the flow of energy and matter in the food chain than he was in the flow of information. In considering an ecosystem as primarily a pathway for the flow of information, not of matter or energy, Rappaport was influenced by Gregory Bateson, another important ecological anthropologist.

Like Rappaport, Bateson had done fieldwork in New Guinea, but much earlier, back in the 1930s. There he met Margaret Mead, later marrying her and doing joint fieldwork with her in Bali. After New Guinea and Bali, Bateson contributed creatively to several disciplines, doing research on topics ranging from schizophrenia to dolphin communication. Both Bateson and Rappaport looked at ecosystems as self-organizing systems of information, drawing on cybernetics. An ecological disaster that causes the breakdown of an ecosystem does finally deplete the flow of energy in the food chain. But just as important is the loss of organization, leading to the breakdown of relationships in the ecosystem.

THE ECOSYSTEM CONCEPT IN ANTHROPOLOGY

So far we have looked only at the concept of ecosystem within cultural anthropology. Eventually this concept became important in all four of the traditional subfields of anthropology, especially in biological anthropology and prehistoric archaeology.

Biological anthropologists adopted the ecosystem concept in the

1960s and 1970s through a set of interdisciplinary research projects called the International Biological Program. These projects studied human adaptation under extreme environmental conditions, including, among others, high mountains (the Andes) and Arctic cold (Alaska, Canada, and Greenland). They turned from the traditional physical anthropologist's measurements of human *structure* (height, weight, and shape) to studying *function*. Functions they studied included the physiological adaptation to cold or to oxygen deprivation at high altitude.

Anthropologists constructed models of the flow of energy through these ecosystems. In the societies most often studied by anthropologists, most of the energy used by humans derived from food and firewood. Since the Industrial Revolution, energy from fossil fuels has become dominant, first in Europe and subsequently almost everywhere else. Even the Inuit of the far north now depend on petroleum, using snowmobiles and boat motors for transportation to their hunting and fishing grounds.

Like biological and cultural anthropologists, prehistoric archaeologists also adopted the ecosystem concept in the 1960s and continue to make substantial use of ecological approaches. They generally are not able to quantify energy flows because of the nature of their data, that is, primarily the trash and other artifacts of past societies. But they are able to study environmental variation over long time spans. These environmental changes are related to changes in human population and settlements, technology, and other features of material culture.

Of the four major subfields of anthropology, anthropological linguists were perhaps least directly influenced by the ecosystem concept. This does not mean that they contributed any less to ecological anthropology. In fact, they played (and continue to play) a very significant part in the development of ecological anthropology through the study of ethnoecology, as was discussed in chapter 2.

Chapter Five

Amazonian Hunters

Beyond the tiny clearings occupied by their thatched houses and manioc gardens, the Achuar occupy a vast area of tropical rain forest in the Upper Amazon. There are about 4,500 Achuar people. Half of their territory lies in southeastern Ecuador, and the other half lies across the border in northern Peru. A French anthropologist, Philippe Descola, began living with the Achuar in 1976, working with his wife Anne-Christine Taylor, who is also an anthropologist. Many other anthropologists initiated fieldwork among other Amazonian peoples at this time, in a veritable explosion of new fieldwork, much of it addressing ecological questions. Rather than reducing all of these ethnographies of Amazonian hunters to a blurred, general picture, we will follow Descola as he describes an Achuar hunting trip (Descola 1994, pp. 222–269; 1996, pp. 120–133).

Pinchu, an Achuar hunter, set out at dawn, carrying his blowgun. He had dreamed about a drinking party at which he and his brother-in-law got drunk on manioc beer and quarreled violently. He considered this dream to be a good omen for encountering a herd of collared peccaries. Indeed, the previous afternoon, while gathering plants near the swamp, he had seen tracks of a herd of about 20 peccaries. Santamik, one of his two wives, came along to carry the food and baggage and, later, the meat. Were they not also accompanied by the nosy anthropologist, the hunting trip would also provide an opportunity for sex in privacy, away from the communal dwelling.

Santamik was also entirely responsible for the highly valued hunting dogs. Their care requires more than simple feeding and training; secret magical songs are considered essential to the dogs' success in hunting.

Ordinarily dogs will hunt armadillos and rodents such as agoutis and pacas. Only the most capable dogs are trusted to hunt dangerous peccaries or to tree an ocelot or jaguar. Dogs are not taken along to hunt arboreal animals (monkeys and birds) or when tracking and stalking game on the ground. The art of stalking is to approach silently, without startling the animal.

Mastery of hunting is above all learning the behavior of the species that are hunted: imitating their calls, knowing their habitat, picking out and killing the dominant male of a troop of howler monkeys first, predicting the response of a wounded animal.

The blowgun is the main hunting weapon of the Achuar. It propels sharp, thin darts tipped with curare, a toxic mixture in which strychnine is the main poison. The Achuar believe that the potency of curare depends on members of the whole household obeying food taboos. Eating sweet foods, especially honey, is thought to weaken the hunter's lung power for using the blowgun, and eating salt might weaken the poison. The silence and accuracy of the Achuar blowgun ensure that it is an efficient weapon, especially in dense forest. The blowgun has not been replaced by the shotgun for hunting game that lives high in the tree canopy, even though all Achuar men now own guns, mostly for use in warfare.

In a typical day's hunting a man would walk 30 to 45 kilometers, starting out on the main trail and then switching to his barely visible hunting tracks to explore an area of up to four square kilometers of steep, muddy, and thorny terrain. This is about a tenth of his hunting grounds.

AMAZONIAN GAME ANIMALS

Descola studied the take for 84 such hunting trips. These trips resulted in 106 kills, a total of 1,200 kilograms (about 2600 pounds) of meat. Despite the fact that the Achuar name some 150 species of animals and birds that they regard as edible game, only 25 species were actually bagged in these 84 hunts. On this particular day, Pinchu first shot a woolly monkey, but it died wedged in a fork of the high tree, unrecoverable by the hunter. Later in the day Pinchu shot two collared peccaries. After butchering them, husband and wife each carried one home, ending an exhausting ten-hour day.

Of the 25 species hunted, only a few species were obtained

repeatedly by Achuar hunters: white-lipped and collared peccaries, woolly monkeys, capuchin monkeys, agoutis, and two kinds of birds—toucans and curassows. The biggest kill was a tapir weighing 242 kilograms (532 pounds), but this animal is normally taboo, so including it in the total distorts the picture.

More than two-thirds of the remaining weight of meat was from peccaries—two species of wild pigs. Collared peccaries weigh about 20 kilograms (45 pounds) and travel in herds of up to ten individuals. White-lipped peccaries are larger—up to 30 kilograms (65 pounds). They travel in large herds of 100 or more animals. Many of the other species of game mammals are difficult to hunt because they are small, arboreal, nocturnal, solitary, or rare. The peccaries, in contrast, are large, terrestrial, diurnal, and abundant (though somewhat unpredictable in occurrence). In light of this, it is not surprising that peccaries are reported as the major game animal for many Amazonian societies.

Anthropologists who studied the Achuar claimed that Achuar hunting was organized in ways that enhanced its sustainability. Descola said that the knowledgeable Achuar hunter "takes care not to shoot wild sows that are pregnant or accompanied by young, in order to preserve the reproductive potential of a peccary horde" (Descola 1994, p. 237). Another anthropologist who did fieldwork with the Achuar living across the border in Peru, Eric Ross (1978), pointed out that the Achuar avoided hunting many larger animals such as the tapir, capybara, sloth, and deer because of food taboos. Some of these taboos were mundane; others were based on the idea that these animals were reincarnated human spirits. Whatever the stated rationale, Ross argued that the species avoided were precisely those most vulnerable to overpredation. Other anthropologists resisted his conclusions, demanding data that would test his assertions about conservation more directly.

As studies of Amazon hunters progressed, many anthropologists were persuaded that the decisions these hunters make about what prey to pursue are most consistent with *optimal foraging theory*. This theory was developed by evolutionary ecologists to describe nonhuman predator–prey relationships, but it seems also to apply for humans, at least in some societies. The theory suggests that hunters maximize their short-term rate of harvest even if this may threaten long-term stability by overhunting a vulnerable species.

One of the studies that supports the optimal foraging model was done by Mike Alvard among the Piro in the Upper Amazon of Peru. The Piro hunted many of the same animals as the Achuar, but the Piro did not hunt with blowguns. Although they are proficient with bow and arrows, they now use shotguns most of the time. During 18 months of fieldwork, Alvard observed 79 Piro shotgun hunts directly as well as interviewing other hunters (Alvard 1995). Here the collared peccary was by far the dominant game animal. The monkeys that the Piro most commonly shot were the spider monkey, howler monkey, and the capuchin.

Alvard found that Piro hunters did not altruistically limit their harvest of easily overhunted species such as tapir and monkeys. Nor did they avoid killing females or adult animals of prime age. Had they been consciously conserving the species, they would have concentrated on hunting mainly the young, the old, or the males of a species. In other words, the Piro were not behaving as "natural conservationists," trying to preserve biodiversity. Nonetheless, this does not mean that the Piro were behaving destructively. Because relatively small numbers of Indians were hunting over very large areas and using low technology, they did not have anywhere near the same destructive impact on game animals as today's loss of habitat to outsiders bulldozing the forest for logging, farming, and ranching.

MANAGING THE FOREST

While some ecological anthropologists in Amazonia concentrated on game animals, others have looked at the relationship between people and plants. They point to clear signs that indigenous people have been managing their forests for a long time. The upland forests of Amazonia contain stands of palms, bamboo, Brazil nut trees, and other useful species that were undoubtedly encouraged by prehistoric farming people that once lived in these areas. These signs of a once more intensively managed environment are one of the clues to the devastating drop in human population throughout Amazonia, after European contact, beginning in the sixteenth century, introduced new diseases and violent exploitation.

Ironically, it was the postcontact decimation of the indigenous population that made it possible for the Indians to do as much hunting as we have been describing. Estimates of the indigenous population of the Amazon vary widely, but we can judge that there were at least three times as many indigenous people in this area when the Europeans arrived as now, and perhaps very many more. Archaeologists exploring old village sites from the centuries just before European contact are finding many more large settlements with extensive earthworks. It would not have been possible to hunt successfully near such large established settlements (Roosevelt 1989).

Wanting to meet the Amazonian people and see their forest environment for myself, I spent my summer vacation from college teaching in 1972 visiting the Culina village of San Bernardo on the border of Peru and Brazil. I was accompanied by my husband, Bill, our two-year-old daughter, Alison, and Patsy Adams, a linguist/missionary/nurse. Eating tapir liver, ceremonially drinking manioc beer, and rocking Alison to sleep in a locally woven hammock, one of my goals was to experience the differences between the New Guinea lowlands and Amazonia. As we flew low over the Amazon basin in a small plane, the endless green seemed tediously uniform, accustomed as we were to the more conspicuously varied terrain of New Guinea from our earlier fieldwork. Underlying the apparent uniformity of all that green is an environmental complexity that anthropologists now appreciate more fully. Early studies made a simple two-way contrast between the *várzea,* the floodplain with soil enriched by sediments washed down from the Andes, and the *terra firme*, the high ground between rivers.

Anthropologists now find that this simple dichotomy needs to be disaggregated, that is, taken apart to make finer distinctions (Moran 1993). The varzea includes the Amazon estuary (affected by Atlantic tides), the lower flood plain with its wealth of fisheries, and the upper floodplain. In the terra firme we need to contrast several ecosystems, too. The lowland savanna has rather poor agricultural potential because of its acid soil and seasonally variable rainfall. Ecosystems in the watershed of blackwater rivers are the most difficult of all. These forests are so low in productivity that hunting is not very successful and people depend more on fishing to supplement their bitter manioc harvest. Upland forests, those outside the flood plain of the Amazon, and montane forests, those in the foothills of the Andes mountains, have soils that are more productive for agriculture. They are the greatest storehouse of the planet's biodiversity but also the most threatened.

THE EVOLUTION OF SOCIAL COMPLEXITY WITHIN AMAZONIA

Just as ecological anthropologists gradually came to understand in more detail the ecological differences within Amazonia, they also came to appreciate differences in political organization throughout the area. The societies that we have so far considered—the Achuar and the Piro—are both egalitarian societies, lacking social stratification. Communities are generally quite small to retain access to good hunting territories. When conflict arises in these societies, their communities tend to fission—part of the group moves away. Because population density is low and resources for farming, fishing, and hunting are fairly uniformly distributed, new settlement sites are not difficult to locate.

In past centuries, larger and more highly differentiated societies arose along the main river. These societies were led by paramount chiefs, political leaders with authority over more than just their small kinship group. These Amazonian chiefdoms collapsed soon after European contact, leaving behind archaeological evidence as well as the accounts of early European travelers (Roosevelt 1989). The rich resources of fish, turtles, and manatees provided the protein resources to sustain larger more sedentary villages well beyond anything that was possible among people of the areas between the main rivers, who relied more on hunting for protein.

In addition to fish, the riverine villages also had alluvial soils, enriched by the sediments from annual floods. These floods were also a hazard, especially for growing manioc, a root crop that requires more than a year to mature. The riverside farmers solved this problem by planting some of their fields on high ground, as insurance against the loss of crops in floods, and by planting quick-maturing maize. Archaeologists note that maize was introduced into the area later, perhaps around 2000 B.C. but maybe as late as 200 B.C.

As population grew along the Amazon, people fought over the choice lands and village sites, as well as for other reasons. When the victors incorporated enemy groups that they had conquered, political groupings became more complex and hierarchical. In short, chiefdoms had evolved.

One of the first anthropologists who articulated this ecological view of political development in Amazonia was Robert Carneiro, who also developed an influential theory of the origin of the state (Carneiro 1970). His view of the origin of the state was similar to his view of the origin of chiefdoms in Amazonia; it emphasized ecological circumscription. Where resources were concentrated in a limited area, warfare and competition over those resources led to conquest and internal stratification. In Amazonia the riverine resources were *somewhat* circumscribed, but in the areas where the first states evolved, resources were *extremely* circumscribed. In the Nile Valley and Mesopotamia, narrow fertile valleys ran through deserts where agriculture was impossible without irrigation. The states that emerged in these circumstances had social inequality and centralized government, as did chiefdoms. But in a state, that central government is territorially based and has much more power than a chief, including the power to collect taxes, draft soldiers or workers, and use violence to enforce laws. The next chapter will introduce some anthropologists who have used ecological approaches in studying state societies.

Chapter Six

Complex Societies

Until now this book has been concerned mostly with hunters and gatherers and tropical forest farmers. These societies are relatively small in scale. A local group of seed gatherers in the Great Basin of western North America typically did not exceed a few families, a total of 25 persons. In New Guinea the local population of Tsembaga Maring was about 200 people. This size is typical of tropical forest farmers. In the Amazon, Achuar settlements of hunters and tropical forest farmers ranged from one house holding a single extended family up to mission-based settlements of as many as 100 people. Each of these settlements was largely self-sufficient economically, except for a few significant items obtained in trade. Several settlements might be grouped into a political unit or alliance of a few thousand people at most.

Studying small populations no larger than those of the Amazon or New Guinea seemed a necessity for ecological anthropologists who hoped to measure the flow of energy and materials through an ecosystem. When ecological anthropologists turned to fieldwork in rural communities that were part of large nations, some of them decided to analyze those communities as ecosystems. They took the relatively small community in which they did their fieldwork as the unit of study, rather than looking at the larger nation or state. For example, Brooke Thomas was one of the anthropologists associated with the Man in the Andes project of the International Biological Program (Thomas 1976). This project took the district of Nuñoa in Peru as its unit of study. The project leaders felt that this decision was justified because the district was isolated both geographically and socially. It was surrounded on three sides by mountain ranges, and most of the 7,750 inhabitants were Quechua Indians with little involvement in the Peruvian national culture. Taking an even smaller unit of study, Thomas presented much of his data for a typical family of six, consisting of parents with four children 2 to 17 years of age.

The people of Nuñoa cultivate potatoes and cereals—quinoa and cañihua. All are traditional Andean crops well adapted to the climate at altitudes above 4,000 meters. They herd alpaca, sheep, and llama. All these animals can cope with the short pasture grasses that grow at high altitudes.

As a human biologist who is also an ecological anthropologist, Thomas was interested in the flow of energy through an ecosystem. All the energy that flowed through the Nuñoan family had its origin from the sun, taken up directly into their crops and herds. They were not using forms of stored solar energy such as gas, oil, and electricity from hydroelectric or nuclear power. The family consumed energy in the form of meat, cereals, and potatoes from their own harvest. This can be measured or estimated as calories. They also expended energy in farming, herding, and other activities that Thomas was able to measure or estimate. He could even estimate the amount of animal dung they needed to collect from their corral for use as fertilizer and fuel.

At the end of all Thomas's measurements and calculations of energy flow through this Andean family, the striking fact was that they were not as isolated as they might initially have appeared. Only about one-quarter of the food consumed by the family came from their own fields and herds. The other three-quarters consisted of foods that they purchased, such as wheat flour, maize, sugar, and alcohol made from sugarcane. All of these high-energy items came from lower altitudes. To get them, the Nuñoans marketed wool, hides, and meat.

Few Nuñoan families at the time of the study owned their own land. They were therefore obliged to do farming and herding work for wealthy owners of large estates, the haciendas. That labor amounted to at least 40 days a year, work that Thomas did not attempt to factor into his calculations of energy flow. In the years following Thomas's study, the people's increasing dependence on the market and wage labor made it increasingly inappropriate to consider the household or the district as an isolated unit. Thomas himself was no longer satisfied with the simplifying assumptions that he had made in order to do his energy flow analysis and recognized the significance of social inequalities and economic change (Thomas 1997). In order to cope with this, ecological anthropologists needed to find new ways to deal with larger units in complex societies.

THE ECOLOGY OF STATES

Beginning with Julian Steward himself, who turned his attention from the Great Basin to the Caribbean region, other ecological anthropologists did their fieldwork in larger populations of thousands or millions. Large societies are politically organized as states rather than bands or tribes. They depend on intensive agriculture, planting grain in terraced and irrigated fields. They consciously shape and engineer their environments to a far greater degree than small societies like the Maring or the Achuar.

Although many anthropologists working in complex societies took small rural communities within them as their units of study, others broke out of this mold. The Norwegian anthropologist Fredrik Barth, for example, focused on larger units. In his 1958 classic paper, "Ecologic Relationships of Ethnic Groups in Swat, North Pakistan," Barth introduced the concept of the *ecological niche* from biology into anthropology.

About 500,000 people lived in Swat State in Pakistan when Barth did his fieldwork there. They comprised three major ethnic groups speaking unrelated languages. By far the largest group was the Pathans, sedentary agriculturalists. They raised wheat, maize, and rice in irrigated fields in the broad fertile valleys of the Indus and Swat Rivers.

The Kohistanis were a smaller group, but with a longer history in Swat. They probably controlled the area from ancient times before being invaded by the Pathans and retreating into their more marginal lands at higher altitudes. The steep terrain required that they construct terraces for their narrow fields of maize and millet. At these higher altitudes, the summer is short. They could only raise one crop of grain each year rather than the two crops that could be raised at lower altitudes.

The third ethnic group in Swat State was the Gujars. Many of the Gujars were true nomadic herders with only sheep and goats, obtaining their grain from the Pathans. They can be thought of as a herding caste in the Pathan caste hierarchy. (In caste systems, people inherit their social position and occupation.) The Gujars in this sense are comparable to other occupational castes that do not farm but provide services to Pathan farmers in exchange for grain. Other Gujars were less fully committed to nomadism. The wives were left to tend the fields and the buffaloes while the men took their herds of sheep and goats to pastures in the high mountains every summer.

In Barth's terms, each of these three ethnic groups occupied a different ecological niche. This is a slightly different use of the term than in biology. In biology, different niches are occupied by different species, whereas Barth applied it to different ethnic groups of the same species, Homo sapiens.

INDONESIAN AGRICULTURE

Just as Barth is known for introducing the term ecological niche into anthropology, Clifford Geertz is noted for introducing the term *ecosystem*. Geertz described "swidden" and "sawah" as two agricultural ecosystems in Indonesia, the fifth most populous country in the world (Geertz 1963).

Sawah (an Indonesian word) is the ecosystem of the flooded paddy rice field. Within Indonesia, sawah is found primarily in densely populated Java and Bali. The irrigation water does more than simply water the rice plants. It regulates the minerals, oxygen, and pH of the soil as well as controlling soil microbes and crop pests of all kinds. The water needs to be kept gently flowing rather than stagnant. Timing is crucial, as the fields need to be drained for planting, weeding, and harvesting, with the water level allowed to rise gradually as the rice plants grow. All of this requires the construction and maintenance of an elaborate system of canals, ditches, and terraced fields, a complete reworking of the natural landscape.

Swidden, now more commonly called slash-and-burn horticulture, is found in the Outer Islands of Indonesia, the sparsely populated islands stretching from Sumatra to New Guinea. Geertz described the swidden plot as a "canny imitation" of the tropical forest. Like the tropical forest it replaces, the field has a high degree of diversity, with a large number of interplanted crop plants. Like the tropical forest, the swidden grows on poor infertile soils; its nutrients are largely locked up in the living plants. The cutting and burning that precede planting are not simply a way of clearing the land but also of making the minerals in the ash readily available for the new crop.

The majority of Geertz's influential book *Agricultural Involution* (1963) is devoted to analyzing Indonesia's colonial history and its implications for agriculture. The production of different export crops was particularly significant: sugar in Java, rubber, coconuts, coffee and other tree crops in the Outer Islands. The pressure of population growth was great on the intensively used lands of Java.

In the decade after Geertz's work, the Green Revolution came to Indonesian agriculture. Indonesia's government adopted an agricultural policy that was intended to increase rice production through the introduction of the newly developed high-yielding "miracle" varieties of rice. In the 1970s the government invested its oil revenues in extending credit to farmers so that they could purchase the fertilizers and pesticides that were required by the new hybrid rice. In order to get two or even three crops per year, the government urged farmers to replant the rice fields as quickly after harvest as possible. They abandoned the planting and irrigation schedules traditionally organized by the water temples.

The new agricultural policies turned sour when the crops were devastated by a series of pests. Chemicals to control pests are not only expensive but become increasingly ineffective as the pests develop resistance and their natural predators are inadvertently eliminated. The first new variety of rice turned out to be susceptible to the brown plant hopper, an insect pest. This variety of rice was replaced by another hybrid seed variety resistant to that insect but vulnerable to a viral plant disease. The next new pests were a fungus disease (Helminthosporium) and rice blast, a disease that causes the plants to wither. Discouraged farmers asked to have the control of irrigation and planting turned back over to the priests of the water temples.

Bali's temples are devoted to a hierarchy of gods and goddesses that ranges from major Hindu gods down to local deities. Their shrines are associated with particular lakes, springs, and canals. The Balinese call their religion Agama Tirtha, the religion of holy water. Sprinkled on or imbibed by worshippers, the holy water is both a blessing—as it comes from upstream—and a purification—as it flows downstream, carrying away impurity (Lansing 1991; Lansing and Kremer 1993).

The congregation of each Balinese water temple consists of people who live and farm in that watershed. For much of the year the temple stands empty, but at least twice a year people gather for festivals and offerings. They also meet to make decisions about what to plant and when to release irrigation water. The decisions they make ensure that the water is distributed fairly and that crop rotation between rice, vegetables, and fallow periods will help to control pests.

Foreign agricultural consultants were skeptical of the idea that religion could have anything to do with pest control. Nor did the anthropologist who set out to study the cultural history of Balinese temples, Stephen Lansing, expect to wind up studying agricultural pests. Lansing did not design his research with the intention of testing Roy Rappaport's theory that ritual regulates environmental relationships—Rappaport's work was not even cited in Lansing's book. Yet it would be hard to imagine a more convincing demonstration of Rappaport's point than these Balinese temples and the rituals associated with them.

VILLAGES IN THE ALPS

The Alps hold a special place in the history of ecological anthropology because two important anthropologists did fieldwork in Alpine villages that helped to move ecological anthropology in a new direction. Each of these anthropologists turned to Europe after he had become well established by major research done elsewhere: Eric Wolf did his earlier work in Latin America, and Robert Netting did his in West Africa. In turning to Europe, both Wolf and Netting chose to work in small rural communities. Anthropologists have been very slow to get into the study of urban ecology.

The village of Törbel is perched high in the Swiss Alps. Villagers support themselves by raising dairy cattle, sheep, and goats in mountain pastures, planting rye and other grains for bread, and gathering wood for housing and heat. There were 125 households in Törbel during Netting's 1970–71 fieldwork. To describe this village in Switzerland Netting borrowed a concept that Eric Wolf had developed in peasant villages of Latin America, that of the closed corporate community. *Closed* in this case is quite literal. In 300 years of parish records only three men had settled in Törbel from outside to marry and raise a family. There was a great deal of out-migration, especially in the late nineteenth century and during and after World War II.

Balancing on an Alp is the revealing title that Netting (1981) gave to his study—"balancing" was very much the theme of the work, showing the stability in this self-sufficient Alpine village. Over the centuries, the village "changed only enough not to have to change." Population growth was restrained by relatively high mortality, late marriage, and frequent celibacy. When gradual population growth did begin to put pressure on food resources in the late nineteenth century, Törbel farmers added potatoes as a new staple food in order to get higher yields from their lands more dependably than they did with grain alone.

Netting continued to call what he was doing "cultural ecology," in the tradition of Steward. His work shared an interest in equilibrium in ecosystems with the newer ecological anthropology of Rappaport and others, though he rejected the measurement of energy flow or the literal application of concepts from general ecology. What was new was his addition of the historical dimension. This was made possible by 300 years of parish records of births, deaths, and marriages. The oldest documents in Törbel's archives recorded land sales as far back as the thirteenth century.

For his Alpine fieldwork, Eric Wolf chose the German-speaking village of St. Felix and the Romance-language-speaking village of Tret in the Tyrol region of northern Italy. Wolf was familiar with the Tyrol, having vacationed there with his family as a boy, later returning when he served with the U.S. mountain troops at the end of World War II. The whole area was part of the Austro-Hungarian Empire until World War I, when it was incorporated into Italy. The villages are only a half hour's walk apart, but they have different ethnic identities. People from the two groups distrust each other and hold negative stereotypes of each other as overorderly versus disorderly, stingy versus spendthrift, and so on.

Wolf did fieldwork in the two villages in 1960–62, work that was followed up by his student John W. Cole in 1965–67 (Cole and Wolf 1974). Their joint research concentrated on inheritance, one of the major cultural differences between the two otherwise very similar villages. The German speakers of St. Felix asserted that the eldest son inherited the land. The residents of Tret favored division of the inheritance. As the study progressed, Wolf and Cole discovered that there was more similarity in the actual practice of inheritance and in ecological adaptation than in contrasting ideas about social structure that people expressed. The researchers concluded that the very real differences in outlook and family structure were not the result of microenvironmental differences, as they had initially hypothesized, but to their historical links to larger cultural entities: the Germanic empire to the north and Mediterranean to the south.

Writing about the Alps, Eric Wolf (1972) was the first one to propose the term *political ecology* for this approach. Political ecology attempts to understand cultural adaptation by taking into account other societies as part of the environment, as well as features of the biophysical environment such as climate and terrain. The term was a play on words, modifying an older term, political economy. As you will see throughout this text, it is rarely possible to understand fully the relationship between a population of humans and some other species without getting into questions of power and inequality. That seems to be true whether the other species are trees in a rain forest, microorganisms that cause disease, or fish threatened by copper polluting a New Guinea river. By the 1980s and 1990s, political ecology came to be the most widely used approach in environmental anthropology.

The next chapter will look at an extended example of political ecology based on the inequality of power and wealth between transnational corporations, the government of a small island nation, and small populations of tropical forest villagers. The mining of a nonrenewable resource—copper and gold ore—devastated the river basin from which the people living downstream from the mine derived their subsistence.

Chapter Seven

The Underground Environment: Minerals

If a tree falls in the forest where no one can hear it, does it make a noise? Another version of this old question is: If there is copper ore in a mountain in New Guinea, is it part of the environment of the people living on that mountainside if they do not know it is there and they lack the technology to extract it? We would probably have to say that it is not part of the "cognized environment," to use Roy Rappaport's term. Nevertheless, as soon as the first mining company geologists enter their lands, the indigenous people are enveloped in a world system that abruptly makes the copper part of their environment. As we will see in this chapter, the significance of a natural resource is relative to the existing technology and economic organization.

Copper was the first metal to be used by humans. This is because native copper (that is, copper in nearly pure metallic form) occurs in pieces that are large enough to be useful much more commonly than native iron, gold, or silver. In most places in the world, including New Guinea, copper is found as a mineral in combination with oxygen or sulfur. One exception is the vast deposits of native copper that were found in northern Michigan and made into tools and ornaments by Native Americans 3,000 years ago, long before Europeans arrived.

Native copper had been used in the Middle East even earlier. By 10,000 years ago, in the early Neolithic period, people hammered and rolled native copper into awls, hooks, beads, and pendants. These early metal workers experimented with heating copper in a charcoal fire and soon developed the process of smelting—extracting metallic copper from copper in mineral form. This extended their supply of copper when the native copper ran out. The simple processes they used required very high-grade ores, oxides and sulfides containing more than 50 percent metal.

Only in the past century has technology been developed that allows miners to exploit large deposits of low-grade ore, that is, rock containing a very small percentage of metal. At the beginning of the twentieth century, steam shovels dug the first open-pit mine at Bingham Canyon in Utah. Nowadays the open-pit copper mines of the western United States, Chile, and New Guinea use huge diesel and electric-powered shovels, bulldozers, front-end loaders, and trucks to dig ore that typically contains about 1 percent copper. In modern mines the metallic copper is extracted from the rock by a combination of chemical and physical processes, including grinding and flotation.

The ability to use low-grade copper ores has one huge environmental implication: For every ton of usable metal extracted, a hundred tons of finely ground rock particles leave the process plant as tailings. Another type of waste around the mine is an equally large amount of overburden, that is, the eroding dirt and rock that was moved to uncover the ore. Both the tailings and the overburden normally contain other heavy metals that were not extracted in the mining process. For many years these wastes were simply dumped around the mines. Sometimes the risk of lawsuits from farmers downstream led mining companies to retain the wastes behind tailings dams. Environmental laws were passed in the 1960s and 1970s in the United States and other industrial countries that restrict the release of mine wastes. However, as we will see in the following case study, the less-industrialized countries are often too desperate for economic development to enforce similar environmental protection laws.

THE OK TEDI MINE, PAPUA NEW GUINEA

One of the ten largest open-pit copper mines in the world is the Ok Tedi Mine in the mountains of Papua New Guinea. Ore from this mine also contains a significant amount of gold, an added source of profit that helped overcome some of the formidable challenges to mining in such a remote and challenging environment. When I visited there in 1983, the mine was still under construction. It had not yet begun to produce copper, but some of its impact on the environment was already visible. Rain forest had been bulldozed to construct the process plant, town site, roads, and other facilities. This earth-moving had produced erosion that made once-clear streams run muddy into the Ok Tedi, the river for which the mine is named.

The Ok Tedi is a tributary of the Fly River, one of the great rivers of the tropics. Because of the heavy rainfall in its headwaters, the Fly is noted for having the highest runoff per unit area of any river in the world. As we motored in a dugout canoe on the Middle Fly with villagers going to their gardens to pick bananas, it was hard to imagine that mining would eventually have a significant impact here, more than 300 miles downstream from the mine.

The Fly River flows through an area that was once linked to northern Australia by a land bridge that has since disappeared. Only the shallow Arafura Sea and the narrow Torres Strait separate the delta of the Fly from Australia. Thousands of Australian birds spend the dry season in the Fly region—tall egrets, glossy ibis, and magpie geese. The birds, like the village fisherfolk, depend on the unique fish fauna of the Fly, fish that are particularly notable for their size. Many of the Fly species are giant species of fish that we know in smaller forms elsewhere in the world; for example, the tiny anchovies that we eat on pizza are related to the giant anchovy of the Fly River, which grows up to 15 inches long. Many Fly River fish species are double the length of species of the same genus found in other rivers of the world.

Developing the Ok Tedi mine in a remote area required that more than one large transnational company be involved. The major investor in Ok Tedi was the largest company in Australia, Broken Hill Proprietary (BHP). The American oil company Amoco was trying to diversify into mining in the 1980s and joined BHP in this venture but pulled out after several years. A group of German companies also invested, trying to ensure a steady supply of copper concentrate for their smelters. The Papua New Guinea government also became a partner, believing that it would do better financially if it held a stake in the mine rather than simply relying on taxes and royalties. (The royalties reflect the fact that, according to Papua New Guinea law, all underground resources are owned by the state, while the land itself is owned collectively by the clans or lineages of each area.)

The Papua New Guinea government was able to borrow money to build a hundred-mile road from the river wharf at Kiunga to the new mine as part of its investment in the project. Kiunga had been a tiny outpost of the Australian colonial government on the Fly River in the 1950s. Now it has become the shipping center where barges unload imported fuel and supplies to be trucked up to the mine. Copper concentrate flows through a pipeline to Kiunga and is loaded on river barges for the trip down the Fly. At the river mouth it is stockpiled to be loaded on oceangoing ships.

In 1984 Mount Fubilan was to be transformed from a 6,880-foot mountain to a big hole in the ground that would be totally mined out by 2009. Fubilan, one of many peaks in the Star Mountain range, had special significance. To the indigenous people, Fubilan was the mythological source of *fubi*, stone axes. Having obtained steel axes in recent years, they no longer needed to trade for stone axes, but they still deemed Fubilan a sacred place because it stood above the underworld, the land of the dead. To the mining company, it was a geological prize because its gold-rich cap could be exploited early on, giving the quick financial return needed to build additional facilities to process the lower-grade copper ore beneath. One of those facilities was a tailings dam to retain the mine wastes that would otherwise enter the river system. Construction of the tailings dam began before the first gold was produced, but work on the dam was repeatedly delayed. In 1989, the government of Papua New Guinea conceded that the mining company would not be required to build a dam.

IMPACT OF THE MINE ON THE LANDOWNERS

Most of the land required for the Ok Tedi mine was owned by the Wopkaimin, an ethnic group of some 700 people. In 1973 an ecologically oriented anthropologist, David Hyndman, began fieldwork with the Wopkaimin (Hyndman 1994). He was especially interested in hunting, the passion of Wopkaimin men. With bows and arrows, they ranged the rain forests from altitudes of 1,500 feet above sea level at the banks of the Ok Tedi to a high plateau at 8,000 feet. Hyndman found that four prey species accounted for most of Wopkaimin hunting success. The most frequently taken game animals were two tree-living marsupials, a cuscus and ringtail. The silky cuscus, also called a phalanger, looks like a teddy bear and weighs less than five pounds (2 kilograms). The coppery ringtail is about the same size. The largest game animals were feral (wild) pigs and cassowaries, which are large flightless birds similar to emus or ostriches.

Hyndman observed that both men and women worked in gardens, growing many crops, of which the staple, taro, contributed two-thirds of the calories in the diet. They planted groves of palm trees for food, including fruit and nut pandanus and starchy sago palms. The women raised pigs. They collected and ate fish, frogs, wild greens, and ferns, contributing to a diverse diet in the years prior to the opening of the mine.

The opening of the mine altered Wopkaimin life drastically, as Hyndman discovered in the 1980s, when he returned from his university position in Australia. Abandoning their rainforest hamlets, the Wopkaimin had moved into crowded roadside villages. They built houses of salvaged packing crates that were later replaced by modern houses. During the construction of the mine between 1982 and 1984, over half of Wopkaimin men worked as unskilled wage laborers, eating in the company dining halls and buying tinned meat and fish for their families to replace the game they no longer hunted. Wives continued to garden, raising sweet potatoes to replace the more demanding taro crop. In this remote area lacking stores and other businesses, beer was one of the few items available for purchase with the wages the men earned. Drinking led to fights, adultery, rape, and other social disorder in the roadside communities.

A dietary study in 1984 revealed major changes in the traditional diet Hyndman had described a decade earlier. In place of the traditional taro, various root crops and store-bought rice had become the staples. Only young working men, working on mine construction and eating in company canteens, had increased their food intake. Child malnutrition remained common. There were new inequities in the availability of food to women, depending on whether their husbands were employed or not.

In the new roadside villages, other health changes took place. Malaria increased with the shift of residence to lower altitudes. Sexually transmitted diseases increased, along with prostitution. In time, as the company opened the health services it had developed for employees to the community at large, there was considerable improvement in community health.

After the peak years of construction, unskilled work for Wopkaimin dropped off sharply from more than 60 percent of the mine's workforce in 1982 to 5 percent in 1986. The skilled workers at the mine came from overseas and from other parts of Papua New Guinea. Some local men found jobs in new businesses surrounding the mine. Others relied on the cash payments for land that was leased by the mine. Some abandoned the roadside villages that were now crowded with squatters from outside the area. They founded new hamlets in the rain forest in order to revive their old male initiation rituals, hunting, and taro farming. All of this took place in a social context drastically altered by the introduction of money into traditional exchanges at marriage, death, and other life crises. Cash income, though not large by Euro-American standards, transformed their subsistence economy and ripped the social fabric.

DOWNSTREAM IMPACT OF THE OK TEDI MINE

The social impact of the mine was undoubtedly greatest among the people who lived closest to the mine. The Wopkaimin lost about 10 percent of their rain forest to the mine. However, an even more significant environmental impact was yet to come to villages downstream that were not expecting it. The Yonggom people in those villages had not even received any income from leasing their land or being employed in construction. The first big sign of trouble, in June of 1984, was a spill of cyanide, the chemical used in extracting gold at the Ok Tedi Mine. Dead fish and crocodiles came floating down the Ok Tedi. The mining company only admitted to the incident two weeks after it occurred.

By 1987 a less toxic but more substantial discharge was having noticeable environmental impact. Mine tailings and waste rock were overflowing the banks and building up in the lower reaches of the Ok Tedi. The fine gray mud cut off the supply of oxygen to the roots of trees, causing their death. By 1997, an area of 100 square miles of riverside forest was already dead, and many more trees were stressed and likely to die soon. Monitoring of fish populations by the mining company scientists confirmed that the Ok Tedi was "biologically dead," as the Australian Conservation Society put it. The Yonggom had lost riverside garden land and sago palms to the flooding (Kirsch 1995). They also found that new and dangerous river currents made it difficult to navigate by canoe. All of these problems would increase in extent every year as long as mining continued. The mining company estimated that eventually almost 400 square miles of forest would die back. Much of the flood plain of the rivers downstream from the mine would be smothered by mine sediments.

Stuart Kirsch, the anthropologist who did research in the villages of the lower Ok Tedi, was not an ecological anthropologist. He had not intended to study the Yonggom subsistence system as Hyndman had studied Wopkaimin hunting. Kirsch was a cultural anthropologist studying the Yonggom symbolic world, particularly their magic and sorcery. Soon Yonggom distress at the environmental damage thrust Kirsch into a different role than he had planned or even than the ecological anthropologists we have talked about in earlier chapters had taken. In this new role, Kirsch found himself an advocate and mediator. His research shifted focus from the village to the global system. The villagers themselves became environmental activists, taking their story to the International Water Tribunal in the Netherlands, the German press, and the United Nations "Earth Summit" in Rio de Janeiro in 1992.

In May 1994 the Yonggom and other downstream villagers took their case to the courts in Melbourne, Australia, the corporate headquarters of BHP. The civil suit against BHP at $4 billion was the largest ever filed in the southern hemisphere. The suit was settled out of court two years later. The downstream villagers had not gotten a huge financial settlement, but along with modest compensation they had extracted a promise that the mining company would build a dam or other means of retaining the mine tailings. By 1997, BHP was ready to try dredging the lower Ok Tedi to put some of the tailings into off-river storage. An unexpected delay in this plan was caused by a severe drought throughout Papua New Guinea, Australia, and Indonesia, associated with the 1997 El Niño, a drastic change in weather due to a shift in Pacific Ocean currents. The dredging equipment was stuck on a barge in the low water of the Fly River. Without water at the mine, copper could not be processed or shipped, and the mine was shut down between April 1997 and February 1998.

A year after mining resumed, the company confessed that the continuing environmental impact was far greater than it had earlier admitted and that BHP did not see any economically feasible solution to the problems created by mining. Even if mining were to cease, wastes would continue to erode into the river system for many years. Furthermore, there were new worries that acid mine drainage would develop, placing further stresses on an already damaged ecosystem.

INDIGENOUS PEOPLE AND ENVIRONMENTAL HUMAN RIGHTS

The failure to build a tailings dam at Ok Tedi showed a reckless disregard for the environment that would not be permitted in the home countries of the investing companies—Australia, Germany, Canada, and the United States. Yet Ok Tedi is by no means the most serious environmental offender among mines. On the same island, New Guinea, but across the border just to the west of Ok Tedi, in the Indonesian province of Irian Jaya, the Ertsberg-Grasberg copper-gold mine was developed by the American company Freeport-McMoRan in collaboration with the Indonesian government. It is now the world's biggest gold producer and third-largest copper producer. Unlike their relatives in Papua New Guinea, the indigenous people of Irian Jaya were not even recognized as owners of the land or compensated for mining. Many local people in the heavily militarized mining area were killed by Indonesian troops.

The volume of ore processed at Erstberg-Grasberg is greater than at Ok Tedi, and the river system into which the wastes erode is a smaller one that has virtually been turned into a tailings dump. In 1990 a broad sheet of sediment flowed from the Aikwa River across to the next river system to the east, killing a three-mile-wide swath of lowland forest in its path and permanently altering the courses of the rivers (Mealey 1996, p. 266).

At both the Ok Tedi and Erstberg-Grasberg mines the environmental rights of indigenous peoples are threatened by transnational mining corporations. The annual sales of large transnational corporations are greater than the gross national product of many of the countries they operate in. Big, powerful companies are not the only players in the mining industry today. There are still small-time gold prospectors much like the forty-niners of the California gold rush 150 years ago. The Amazonian gold rush of the 1980s and 1990s was largely composed of these small operators, aided by helicopter and small-plane transport. Most small miners use methods that are highly polluting, contaminating water and fish with mercury. The miners have brought epidemics of malaria and other diseases to the indigenous people, along with the violence and social disruption common to frontier towns. Anthropologists have stood with indigenous leaders in calling attention to the abuses caused by mining (Sponsel 1997).

Often it is one or two anthropologists who serve as advocates for the rights of the people they have studied. Sometimes they bring their concerns to a wider group of anthropologists, such as the American Anthropological Association (AAA), asking their colleagues to pass resolutions speaking out against violations of human rights. The AAA has a permanent, elected committee working on issues of human rights. Nongovernment organizations, such as Amnesty International, are concerned with human rights generally, but in order to focus more specifically on the rights of indigenous peoples, an anthropologist at Harvard University, David Maybury-Lewis, founded the organization Cultural Survival. Through its magazine, *Cultural Survival Quarterly* and its Web site—www.cs.org—Cultural Survival educates the public and policy makers about threats to indigenous peoples and ethnic minorities. Similar activities are undertaken by the British-based organization Survival International and the International Work Group for Indigenous Affairs. Indigenous organizations also play an essential part in keeping a spotlight on their struggles for human rights (Bodley 1999, Appendix A).

Many of the violations of basic human rights of indigenous people have a basis in conflict over natural resources. Historically, ethnic minorities were pushed into areas unsuitable for farming and industry. The discovery of oil or minerals and the need for timber in the industrialized world abruptly removed the protection their isolation once afforded them. Indigenous people who live at great distances from the centers of power and wealth share something with the most impoverished people in industrial societies. Both of these groups are asked to bear a disproportionate share of the *risk* associated with industry while at the same time receiving a smaller share of its *benefits*. The following chapter will consider the concept of risk in more detail.

Chapter Eight

Hazard and Risk

Most of ecological anthropology deals with everyday life, the times when things are going along smoothly. The very use of terms such as *adaptation* and *equilibrium* hints at an emphasis on average or normal conditions. Yet it may be that the way that individuals and groups deal with extremes tells us the most about adaptation.

The types of hazards that endanger people, threatening their survival, are often seen as being divided into *natural* hazards, such as drought, flood, or earthquake, and *social* hazards, such as war or civil unrest. One of the contributions of environmental anthropology is to complicate this oversimplified distinction, that is, to show how natural hazards are always relative to the cultural context in which they occur while social hazards have causes and consequences in the natural environment.

A disaster is an extreme manifestation of risk, concentrated in time or place. The boundaries between disasters and ordinary, recurrent forms of risk are fuzzy. We perceive a plane crash with 130 lives lost as a disaster but do not call 430,000 deaths per year in the United States from cigarette smoking a disaster. The deaths from smoking come through long-drawn-out chronic diseases and are spread out in time and space. Even so, smoking is much riskier behavior than flying, in statistical terms.

Most disasters are located somewhere in the middle of a continuum between the natural and the social. The death of 500,000 people in 1988 in a famine in Sudan had a natural component in drought, but the shortage of food in the rebellious southern and western parts of the country was largely engineered by the Khartoum government for political and economic purposes. Had it chosen to do so, the government could have prevented many deaths by allowing available food to be distributed (Keen 1994). The death tolls from the 1999 earthquakes in Turkey were increased because of lax enforcement of building codes in the construction of the apartment buildings in which victims were crushed or trapped. "Natural" disasters such as drought and earthquake are thus inextricably bound up with the social and political.

Technological disasters similarly combine the social and natural. The deaths from radiation in the 1986 explosion at the Chernobyl nuclear power plant resulted from the deployment of a risky technology of nuclear power in the Soviet Union by a political-economic regime that did not exercise adequate safeguards. The elevated levels of thyroid cancer, leukemia, and other diseases that continue to occur in children over a wide area in the following years are combined in health statistics of the same disease from "natural" causes. It is not possible to say of any one child that this child's disease is not one of the much smaller number that would have occurred anyway, without the exposure to radiation from the explosion.

In environmental anthropology, we are interested in how societies manage the hazards to which they are prone. How do high-altitude farmers in New Guinea prepare their fields and diversify their plantings for the occasional but infrequent frost? How do Americans establish the locations for hazardous waste dumps? How do the livestock herders of East Africa manage to get through extended times of drought?

WATER: TOO LITTLE AND TOO MUCH

The Ariaal pastoralists of northern Kenya are one such group who were studied through good times and bad by an environmental anthropologist, Elliot Fratkin. The Ariaal are a population of about 10,000. They are a mixture of two cultural traditions—the Rendille, speaking a Cushitic language related to Somali, and the Samburu, speaking a Nilotic language related to the language of the Maasai. All of these groups are herders of cattle and other livestock. Fratkin began his study of the Ariaal in 1971 and has gone back to Kenya several times (Fratkin et al. 1998).

Fratkin observed that an average household of 5 people kept 12 camels, 20 cattle, and 50 sheep and goats. The herds provide milk, 70 percent of the Ariaal diet on a daily basis, and occasionally meat and blood for food. The animals are also sold to purchase grain, sugar, and tea—other essentials of the diet. The different kinds of animals in the herd behave differently, with different strengths and weaknesses, making it important to keep a diversity of stock. Goats and camels browse on twigs and leaves, while sheep and cattle prefer to eat grass. Camels are good milk producers but have a low population growth rate because of a low number of births as well as the death of their young to infectious disease and parasites. Sheep and goats reproduce much more quickly, so they are important for building up a herd after a drought and providing meat and cash. Cattle are most vulnerable to drought, which has returned every four or five years since 1968. During the long drought that lasted from 1982 to 1984 the Ariaal lost more than 50 percent of their cattle.

The Ariaal strategy for coping with hazards is to keep large numbers of different stock, to maximize herd size, and to disperse their herds in different areas. They can do this by lending animals to friends and kin far away, where conditions may be somewhat different. Mobility used to be an important strategy, but it has been lost with the loss of grazing land under pressure from population growth, the expansion of agriculture, and the creation of commercial ranches and game parks. Under severe conditions, such as those experienced in 1984, the usual mechanisms for coping with drought may not be enough. The most impoverished Ariaal families were driven out of pastoralism and into town, where they depended on famine relief agencies or odd jobs. The luckiest were able to benefit from other opportunities offered in town: education, health care, and income earning through market gardening or other enterprises.

Like the scarcity of water, the overabundance of water is a culturally mediated hazard. In some times and places, a flood may be a disaster, in other times and places it may be a resource. The flood of 1988 in Bangladesh was by all accounts a disaster. The Buriganga River suddenly flooded most of the city of Dhaka (Shaw 1992). Despite fears of looting if they should leave their homes, thousands of people could no longer live on the roofs of their flooded houses, so they entered relief camps.

A severe flood in Bangladesh has more serious implications for women than men because of the requirement that they maintain purdah. Ideally this means that they remain secluded within the household, but they cover themselves if they must go out. Muslim Bengali women are also defined as subject to pollution through menstruation and childbirth. One midwife detailed the serious difficulties she faced when attending the birth of her grandchild, born during flood time in a camp. She was unable to conduct the rituals of seclusion or burying the placenta and cord under the hearth of the family home. These ritual failures are considered dangerous to the health of the child as well as to the cattle and crops (Shaw 1992).

This flood did not affect everyone equally. Bangladesh depends on its annual floods to create and restore its fertile alluvial soils. The alluvial soil is earth eroded high in the Himalayas and deposited all across the vast river delta that covers most of Bangladesh. The agricultural system is adapted to the expectation of flooding. Farmers who are able to do so plant both flood- and drought-tolerant varieties of rice at different elevations and seasons in order to be ready for any eventuality. Poor farmers are less likely to have the land and labor available to do this. The poorest people suffer most, particularly destitute women abandoned by their husbands.

RISK AND ARCTIC HUNTING

Hazards are sometimes referred to as risks to emphasize that there is an element of chance, or probability, associated with their occurring in a given period of time. This probability can be expressed mathematically, which is a more precise use of the term risk. Engineers speak of designing a dam for a "fifty-year flood," meaning that the magnitude of a flood that will overtop the dam can be expected to occur, on average, once every fifty years. (Of course, they cannot tell you whether that will be next year or the year 2050!) Insurance companies and gamblers are also concerned with the probability of certain events. Environmental anthropologists who use optimal foraging theory to study hunting and gathering, as mentioned earlier in the chapter on the Amazon, also talk about risk in quantitative terms.

The Inujjuamuit of the Canadian Arctic were foragers who traditionally hunted for caribou in winter, trapped foxes, set fishnets, hunted by canoe for seals, and hunted birds (ptarmigan and geese), among many other activities, which continue today though not as frequently as in the past. For each type of Inujjuamuit hunt there is an average rate of return. The anthropologist Eric Smith measured this rate of return in kilocalories per hour. (Alternatively, he could have expressed the return in other units such as weight of meat or grams of fat and protein.)

Impressive returns came from the winter caribou hunt (6,690 kcal/hunter-hour). According to Smith's calculations, it was also efficient to set out fishnets under lake ice to catch whitefish, lake trout, and char (7,140 kcal/hunter-hour). Some hunts had a much poorer return: Hunting ptarmigan with a .22 rifle from a canoe returned only 240 kcal/hunter-hour (Smith 1991, Table 7.5, p. 272). This kind of ptarmigan hunting was only done in the fall when the birds gathered in large flocks, the weather was unsuitable for marine hunting, and the hunters were traveling to a spot where it was also convenient to set out fishnets.

Different Inujjuamuit hunts that offer a similar average rate of return may still show different amounts of variance around that average. This is where risk comes in. If there is a high degree of variance, it means there is a chance of coming home at the end of one day with nothing at all and another day with a very large catch. Different hunters have different attitudes toward this risk.

One characteristic of studies like Eric Alden Smith's study of Arctic foraging is that they allow us to convert all kinds of activities to comparable units of time, calories, and money. This has the advantage of giving some insight into decisions that Inujjuamuit must make. Of course it leaves out cultural values such as aesthetic, spiritual, and social aspects of hunting that cannot be reduced to a universal standard of value. Inujjuamuit can readily earn money by selling their soapstone carvings. This is especially true for older men who are highly skilled after many years of practice (Ann McElroy, personal communication). With the money, they can purchase imported meat and fish at the store for an average of $5.97/kilogram. For even better value for their money they can buy "country food," caribou meat, geese, beluga, and local fish, at the co-op for $2.14/kilogram. The foods they get by spending their time hunting and fishing "cost" them $7.34/kilogram if the time lost from carving is counted along with the direct costs such as fuel and ammunition. Inuit, like the rest of us, are not making decisions on a purely monetary basis.

ASSESSING RISK

In societies with technological specializations such as engineering and insurance underwriting, laypeople evaluate risk differently than specialists in risk analysis. Laypersons take into account not only the size and probability of risk, but questions of fairness, equity, and a sense of control. Cultural differences lead them to select some dangers for public concern while downplaying others that are taken more seriously in another culture. Furthermore, people do not often have all the information they need to make rational, informed decisions because powerful institutions such as corporations and governments control such knowledge and use it to manage people's responses. This was the case at the Ok Tedi Mine, where the mining company published a glossy public relations brochure, reassuring people that the mine was environmentally harmless, only to reverse itself when it was no longer possible to cover up the environmental damage.

Public health officials and other government bureaucrats are often in this position because they want to reduce public anxiety and panic. They issue reassuring statements until they can clarify what the dangers are. These reassurances sometimes deny people information they need to make good decisions. Anthropologist Gregory Button (Button 1995) discovered this in studying the public health response to a Shetland Islands oil spill in 1993 in the North Sea, north of Scotland and west of Norway. Button had earlier studied the Exxon *Valdez* oil spill in Alaska. The Shetland Islands spill, though less publicized than the Alaskan one, involved twice as much oil, which spilled when an American-owned oil tanker went aground. The spraying of toxic chemical dispersants that were used to clean up the oil spill unfortunately contaminated agricultural land, sheep, cattle, and people. The residents were given inadequate warning and information, and their health was not adequately monitored. Had they known more, they could at least have stayed indoors and reduced their exposure to the toxic sprays.

People in all cultures have to make many important decisions with inadequate information about the environment. The tropical forest is a particularly complex system, so the processes of decision making in tropical forest farming are challenging. There are too many variables of soil, water, vegetation, pests, and weather to predict reliably the outcome of particular plantings. Under some circumstances the best decision, the decision that is least likely to lead to a crop failure from floods, dry spells, pests, or other disasters, is to flip a coin, that is, to randomize. This is the situation in the selection of new sites for swiddens in Borneo, argues Michael Dove (Dove 1993). Dove studied the process by which the Kantu' of West Kalimantan (Indonesian Borneo) decide where to plant their new fields. As they set out to clear a site, they look and listen for omens that indicate the likely success of the field. They use bird augury, a form of divination in which the calls and flight patterns of certain birds reveal messages from benevolent deities. The way the omens are interpreted virtually assures that the farmers will diversify and randomize their choices rather than using flawed assumptions based on the previous year's conditions.

The hazards faced by farmers in Borneo and Bangladesh, hunters in the Arctic, and pastoralists in Kenya, while differing in specifics of climate, share a common theme of scarcity. All these people faced the risk of going hungry if they made bad decisions. In our own late industrial society, the dangers we face are less often those of scarcity than ones produced by industrialization itself: pollutants and toxins. Some of these are localized in the community downriver or downwind of an abandoned mine, chemical dump, or nuclear research site. Exposure to such risk is greatest among the poor and powerless. Other dangers are as global as the fallout from a nuclear explosion or the climate change resulting from the increased production of carbon dioxide and other greenhouse gases. These threaten all of us, though the distribution of risk across the world population is uneven.

Chapter Nine

Population

On a day in October 1999 the population of the world passed 6 billion. There were signs that the rate of reproduction had leveled off slightly, but the momentum of children already born growing to adulthood and having children of their own assured that the population would not stop growing anytime soon.

Anthropologists are trained to take the long view on the big questions. World population is one of those big questions. It had taken only 40 years for world population to double, going from 3 billion in 1960 to 6 billion in 1999. Back in 1850, in the middle of the Industrial Revolution, the world population was only 1 billion. Reaching back farther, to the beginning of agriculture in the Middle East about 10,000 years ago, world population was probably somewhere between 5 and 10 million—only about 0.1 percent of the population at the end of the twentieth century.

This long sweep of human population history forms the background for all of anthropology and its subdisciplines. Prehistoric archaeologists, for instance, have attempted to identify evidence left behind by tiny populations of hunter-gatherers soon after they entered new continents, such as Australia and North America. Cultural anthropologists and biological anthropologists studying contemporary hunter-gatherers have cautiously used them as models of population processes of fertility and mortality in ancient foraging societies. Linguistic anthropologists have tracked the spread of the Polynesians into the islands of the Pacific by showing how the related languages there have diverged over the past several centuries. Applied anthropologists, working as consultants in family planning programs, have investigated what motivates families in Indonesia or India to accept or reject contraceptives. Population is relevant in a wide variety of studies by anthropologists.

POPULATION AND ENVIRONMENT IN THE MAYA LOWLANDS

One of the great puzzles of archaeology is the buildup and subsequent collapse of the great Maya civilization that occupied the lowland tropical forests stretching from Mexico's Yucatan Peninsula through Belize and Guatemala to Honduras. Impressive ceremonial centers such as Tikal and Copán reached their peak in the Late Classic period between A.D. 700 and 800. The central authority collapsed and the elaborate structures fell into disuse long before the Spanish conquerors arrived.

The rise and fall of Maya civilization can be tracked as a rise and fall of population. One of the best-studied of the Classic Maya centers is Copán in western Honduras. The urban center of Copán contained a palace complex where the kind and nobles lived and an elaborate collection of pyramids, temples, tombs, and sculptures. At its peak population, between A.D. 750 and 900, the city may have had as many as 12,000 people living in one square kilometer. A similar number lived in settlements scattered throughout the Copán valley. From its peak totaling about 26,000 in A.D. 900 the population began dropping. By A.D. 1000 it had diminished to about 7,500 people, and by A.D. 1200 to about 1,000 people. Soon after this, the area was virtually abandoned. As dramatic as this crash sounds, until recent studies were complete, archaeologists thought the peak population was even larger and the crash even more abrupt (Webster, 1999; Webster, Freter and Gonlin 2000).

How do archaeologists know what the population of these ancient Mayan centers was in different eras? The Mayans had hieroglyphic writing but did not, as far as we know, keep censuses. The main clue to population size is counting the number and size of structures occupied in a given phase. The phases of occupation are usually dated by ceramics. The house-count method requires making an assumption about the average number of persons per household, typically taken to be between four and six persons for a nuclear family. These methods are indirect enough that controversy remains about the extent of population decline. In some Mayan areas outside the Copán valley, there does not appear to have been a population decline so much as a scattering of people out over the countryside away from the urban centers.

The prevailing explanation for the Maya "collapse" at Copán has been an ecological one. According to this model, the downfall came about because the people attempted to live on maize as a staple food in a tropical lowland environment under increasing population pressure. Deforestation, the invasion of grasses, soil depletion, and erosion cut into productivity, ultimately causing nutrition and health to deteriorate. People became more vulnerable to infectious and parasitic diseases, especially diarrhea, and the dense settlement helped these diseases to spread. This led to dispersal and depopulation.

At the time of the conquest the Spanish did observe that slash-and-burn single-crop maize fields were the only form of subsistence in the severely depopulated Mayan region. But archaeologists now know that at the peak of population, the Maya were practicing more varied, sophisticated, and intensive farming. They built raised fields in wetlands and constructed terraces to manage water and soil fertility. More recent studies refer to this pattern as a "managed mosaic" (Fedick 1996). This does not mean that the Maya may not still have degraded the environment, although it does give us a new respect for their agricultural knowledge.

The "dirt" archaeologists who have uncovered house platforms and pots are not the only anthropologists who contribute to an understanding of the population and environment of the Mayans. Bioarchaeologists study the bones and teeth in Maya burials. The Maya buried their dead under the floors of individual houses. With no big cemeteries to give large samples of skeletons, it is a somewhat slow process to build up a picture of changing health conditions across time, place, and social status. Some of this skeletal evidence does give support to the ecological hypothesis. Comparing the length of leg bones, for example, suggests that, on average, men were shorter in later time periods. As population grew, food may have become scarce, stunting growth.

Defects in the enamel of children's teeth also support the conclusion that Maya health deteriorated over time. The teeth show that children experienced significant nutritional stress around age three, the usual age of weaning. In the Late Classic period at Copán, Rebecca Storey (1992) found that even children buried in a relatively wealthy elite residential group were under nutritional and disease stress. Not all of the Mayan bioarchaeological evidence clearly supports the ecological hypothesis, though. Diet and nutrition were variable across the region (Whittington and Reed 1997). Currently there is a great deal of controversy about how best to interpret the signs of malnutrition in ancient Mayan skeletons (Wright and Chew 1999).

Few anthropologists find the ecological explanation for the Maya decline by itself fully satisfying, and they are adding a political dimension to their explanations. In the Mayan case, at Copán, for example, it is clear that the last kings were losing their grip on their kingdom. Privileged people, the nobles, were increasingly powerful and continued to have access to valuable imported goods such as pottery for deposit in their own tombs, even after the construction of royal monuments ceased. The royal power collapsed abruptly, but it is not clear whether the final blow to it came from internal political conflict or external war. There is little indication from trauma to skeletons or the sudden destruction of buildings that there was mass violence at the time of the so-called "collapse" (Webster 1999).

THE ANTHROPOLOGY OF DEMOGRAPHY

Unlike archaeologists, cultural anthropologists in the field can and routinely do take a census of the communities in which they do their fieldwork. Except when going house to house and asking questions might create the suspicion that the newly arrived anthropologist is really a tax collector, this is usually one of the first things that a new fieldworker does, in order to get acquainted with the neighborhood. I found that I needed to take a census every day in my small New Guinea village because people were constantly coming and going.

While an ethnographer's census can be highly accurate, the small size of the population studied usually rules out using the statistical methods that ordinary demographers have developed. Fortunately, some anthropological researchers have specialized in developing methods for working on the demography of small populations. Nancy Howell (1979) brought a sociologist's methodological sophistication into the study of an anthropological population, the !Kung San of southern Africa. She examined particularly the question of fertility and the long spacing between births among nomadic hunter-gatherers. John Early and Tom Headland (1998) analyzed the population data that Tom and Janet Headland had accumulated over forty years with the Agta of the Philippines. They showed how the Agta hunter-gatherers were being assimilated into lowland farming society by migration and intermarriage. Both of these studies deal with small and relatively isolated societies.

One of the most thoroughly investigated populations of forest foragers are the Northern Aché of Paraguay. There are now about 590 individuals. This compares with about 540 in 1970, just before the Northern Aché came out into peaceful contact with the outside world at mission settlements after decades of violent conflict with peasant colonists. Immediately after contact, the population declined by one-third to one-half because of epidemic diseases to which the Aché did not have immunity. The population was able to rebound quickly because so many of the survivors were young adults (Hill and Hurtado 1996).

Anthropologists who study villages or neighborhoods within large or complex agricultural or industrial societies make a different kind of contribution to the study of population. They leave the highly quantitative demographic research to be done by the interviewers and demographers working for the national population census with their large-scale survey methods. Anthropologists emphasize the face-to-face, open-ended interviews and participant observation techniques that they traditionally do best. These intensive methods give insight into how people establish marriages or sexual partnerships, attitudes toward contraceptives, and other beliefs and behaviors underlying population processes. They document the value of work done by children in herding animals, gathering firewood, and babysitting younger children, especially in intensive agricultural societies. Fertility is likely to remain high as long as the value of children's labor is high and the cost of raising them is relatively low. As this ratio changes, families come to be satisfied with fewer children.

The same question about the value of child labor in relation to fertility can be asked of historical data. For example, population growth in Ireland was at its highest level in the late eighteenth century. This was the same period when the English textile industry had a high demand for imported Irish linen thread. Flax harvesting and spinning increased the value of child labor, especially work by young girls (Ross 1986).

DEVELOPMENT WITHOUT GROWTH IN KERALA, INDIA

Not all low-income societies practicing intensive agriculture show equally high fertility and mortality. An intriguing exception is the Indian state of Kerala. Kerala is a small strip along the subtropical southwest coast of India. Kerala's population is a tiny portion (a little over 3 percent) of all India's population (1 billion in 1999). Yet, to put this in perspective, Kerala has approximately the population of Canada (30 million), while occupying approximately the land area of Switzerland. Although the population of Kerala is very dense, it is somewhat evenly distributed over the land, without sharp differences between cities, towns, and villages. Nor are there sharp ecological differences in resources. The land and water necessary for wet-rice agriculture are uniformly distributed. This ecological uniformity gave Kerala a head start in providing its people with equal access to services such as roads, schools, and health centers (Franke and Chasin 1994).

By various measures of demography, Kerala is more similar to a wealthy industrialized country than to the rest of India or to low-income countries. Life expectancy (in 1991) is 69 years for men and 72 years for women, compared with 76 years in the United States and 60 in all of India. The higher life expectancy in part reflects low infant mortality: 17 (per 1,000 live births) in Kerala versus 9 in the United States and 85 in all of India. The birth rate in Kerala is 20 (annual births per 1,000 people), compared with 16 in the United States, 31 in all of India, and 38 in the whole group of poor countries, those with per capita incomes below $635.

These demographic indicators for Kerala would not be so surprising if it were a relatively wealthy part of India. Quite the contrary, Kerala is one of the poorest parts of India. Its per capita income of $298 (in 1991) is below the average of $330 for India as a whole and is comparable to that of the poorest countries in the world. Unemployment is high.

What is different about Kerala that could account for its achievement of low fertility and low mortality? These are the demographic indicators of a high quality of life, but Kerala has achieved them without the high incomes that are usually thought to be a precondition. Literacy is also high in Kerala. What is especially important for child survival and low fertility is that literacy among Kerala women is almost as high as that for Kerala men, in contrast with the big gap between genders in the rest of India. Richard Franke and Barbara Chasin (1994), anthropologists who did fieldwork in Kerala in 1986–87, concluded that the interconnected factors in Kerala's successful development were ecology, a long history of international trade, and progressive organizations. Dating back nearly a century, Kerala has a local tradition of workers' and small farmers' organizations. Well in advance of other areas, these organizations in Kerala pushed for the successful establishment of programs of redistribution in areas such as school lunches, food rationing, public housing, accessible health services, and land reform.

POPULATION GROWTH AND ENVIRONMENTAL DEGRADATION IN HONDURAS

Environmentalists make a simple and powerful argument about the relationship between population growth and environmental impact. It is summed up in the equation

Impact = Population x Affluence x Technology. The IPAT equation was popularized in the book *The Population Explosion* (Ehrlich and Ehrlich 1990). This makes sense in relation to the United States, for example, with the third-largest population of any nation, great affluence, and technology that is heavily based on the internal combustion engine. This combination of population size and consumption means that the United States has a huge impact on the global environment.

Applying the IPAT equation to the developing world suggests

that if more people succeed in their aspiration to live at the same level of affluence as in the United States, the impact on the environment will be severe. Unless the developing countries slow their population growth and choose less damaging technologies, the quality of life for everyone will suffer. The IPAT equation is a good commonsense argument. Nevertheless, it is one that anthropologists have found inadequate because it oversimplifies the processes at work in the human impact on the environment (Durham 1995).

The Central American countries provide a clear instance of environmental degradation that cannot be understood simply as a direct result of local population growth multiplied by affluence and technology. This is so even in Honduras, the most rapidly growing country in the region, a mountainous tropical country of 5 million people with a growth rate of about 3.3 percent, making it one of the most rapidly growing countries in the world.

The forests of Honduras were cut to create pastures to feed cattle during the beef boom of the 1960s and 1970s. The beef was largely exported rather than fed to the growing local population. When the cattle boom ended in the early 1980s, partly because of U.S. restrictions on beef imports, the government encouraged the development of nontraditional export foods, especially melons and shrimp. Both these new exports were environmentally damaging. Melon farmers used high levels of pesticides. Shrimp farmers cultivated shrimp in large ponds developed in coastal wetlands, modifying coastal mudflats and mangrove ecosystems. Road building and the buildup of sediments caused further problems. The private development of shrimp farms cut off public access to fishing grounds, reducing an important source of food for local people (Stonich 1993).

These environmental impacts are not a simple result of population growth. Anthropologists such as Susan Stonich and William Durham, who seek to understand the degradation of the environment in Central America, point to the inequality of access to resources as the basic cause of environmental degradation. The rich are encouraged by market demand and capital accumulation to expand the commercial production of beef cattle, cotton, melons, and other export products. This leads to deforestation and to the further concentration of land ownership when they buy up more land with their profits. The poor are pushed off these lands. Further impoverished, they intensify their own household production, using marginal lands such as those higher up the mountain slopes. These marginal lands then become deforested and eroded. Erosion exposes the slopes to the increased likelihood of landslides during the heavy rains that come with hurricanes. The rich get richer and the poor get poorer in two positive feedback loops, both of which lead to more deforestation.

While Stonich and Durham regard inequality of access as a much more fundamental problem in Honduras than population growth, this is not to say that population growth is unrelated to environmental degradation. Indeed population growth is encouraged by Honduran poverty. Larger families provide child labor in poor households and cheap labor for the new developments.

One set of side effects of Honduran environmental degradation is the impact on human health. The indiscriminate use of agricultural pesticides on fields of export crops destroyed the predators of the whitefly, which transmits a virus disease of beans. Damage to the bean crop, on which the poor depend for protein, led to increases in child malnutrition. The use of agricultural pesticides also led to widespread resistance to insecticides by anopheles mosquitoes, spreading malaria in populations that lacked immunity to it. Public health specialists have found that environmental change in Honduras has affected the insect vectors that carry other diseases, increasing the rates of leishmaniasis and Chagas' disease, both of which are tropical diseases caused by protozoa (Almendares et al. 1993). In the next chapter we will look at other ways in which the loss of biodiversity affects human health.

Chapter Ten

Biodiversity and Health

The word *biodiversity* most often refers to the abundance of species; loss of biodiversity has to do with the extinction of species. Scientists have described about 1.4 million species; that is, they have assigned them a proper Latin name that differentiates them from all the other species. At least double that number still remains to be described. No one knows how many species of insects, fungi, or microorganisms there may be. Even among the best studied groups—the birds and mammals—new species continue to be discovered, though it becomes increasingly likely that they become extinct before being described as their habitats are lost through deforestation. The loss of the insects, birds, and bats that have specialized relationships with plants for pollination and seed dispersal is particularly worrisome, for it would cause a cascade of other extinctions. The World Conservation Union (IUCN) estimates that over 34,000 species of plants are threatened with extinction, some 12.5 percent of the world's flora. The well-known Harvard biologist E. O. Wilson estimates that the rate of the loss of species has accelerated so that 10,000 times as many species disappear each year now as disappeared per year before humans appeared on the scene. The cutting of rain forest is the biggest cause of lost species (Wilson 1989).

The species level is not the only level at which biodiversity is important. Above the species level, there is a diversity of ecosystems or landscapes, larger units that can collapse or disappear. Below the species level, there is genetic diversity within each species to be preserved or lost. Some species are much more diverse than others are; for example, the other primate species are much more genetically diverse than the species Homo sapiens.

Humans have often fostered biodiversity for the sheer pleasure of it. Think of the many different breeds of dogs of the one species *Canis familiaris*. Some of them were bred for practical uses such as herding sheep, hunting birds, guarding against intruders, or racing. Many breeds today are raised for their appearance and for prestige.

People value diversity in plants, too. Like many home gardeners, Celia Ehrlich grew the Hawaiian ti plant (*Cordyline fruticosa*) as merely an attractive tropical houseplant. Then in an anthropology course she read Rappaport's *Pigs for the Ancestors*. The book described how the Tsembaga Maring ritually uprooted a ti plant at a critical point in the ritual cycle of pig sacrifice and war. (In the Tsembaga language the ti plant was called *rumbim*.) By exhaustively searching the literature, ethnobotanist Ehrlich found that throughout most of Southeast Asia and the Pacific, people traditionally planted ti for ritual use. They planted a huge number of ornamental varieties with different colors of leaves from green to yellow to red or purple. The plant has a starchy edible root, but only in a small part of the region did people bother eating it. Instead it was the symbolic and ritual use that was important (Ehrlich 1999).

One particularly important kind of biodiversity is the variation within species of crop plants. Since the origin of agriculture some 10,000 years ago, farmers have carefully maintained at least small plantings of many varieties of important food crops such as rice or potatoes. Each variety has its own special qualities, such as resistance to a particular pest, different flavor or texture, or perhaps simply an attractive leaf color. Different varieties thrive in slightly different microenvironments under specific conditions of soil, water, and temperature. The Green Revolution beginning in the 1960s encouraged the planting of new high-yielding crop varieties that depended on fertilizers, pesticides, and irrigation. Since the development and promotion of these varieties, there has been good reason to worry that the genetic diversity of crop plants will be lost. If this diversity is lost, we will not have the genetic materials that will be needed to breed improved plants for future needs, as insurance against climate changes or new diseases.

An anthropologist doing research on agricultural decision making in the Philippines, Virginia Nazarea, became concerned that a great deal of agricultural knowledge connected to these older varieties of crop plants would soon die out. Working with rice farmers in Budiknon province on the island of Mindanao, she began to document the indigenous beliefs and practices associated with sweet potato cultivation. She called this "memory banking," in contrast to gene banking that preserves only the genetic variability. As an extension of her project, a plan was developed to maintain an "in situ germplasm collection," that is, to keep the old varieties going in local gardens and not only at agricultural research stations (Nazarea, 1998).

THREATS TO BIODIVERSITY

Despite the pleasure that humans take in diversity, we are also the most important threat to biodiversity. Hunters may have been partly responsible for the extinction of large animal species throughout the world toward the end of the Ice Age (Martin and Klein 1984). Some scientists doubt that the extinction of mammals such as the mammoth and giant beaver was due purely to climate change because the extinctions occurred at different times in different continents after human hunters entered the landscape. More recently, as humans settled the islands of the Pacific, they caused a wave of extinctions of bird species. The birds succumbed not only to being killed by people but, more often, by the animals that people brought along in their boats, including dogs, rats, and snakes.

Despite the dangers posed by people, it is rare that humans drive other species to extinction through directly hunting down their last representatives. The biggest threats are more indirect: the destruction of habitats, the fragmentation of habitats, and the introduction of exotic species that compete with native species.

When the loss of biodiversity threatens the food supply, it has obvious implications for human nutrition and health. Even after the development of agriculture and the domestication of animals, people throughout the world depend on wild plants and animals as a source of variety in their diet and as backup in times of famine.

MEDICINAL PLANTS

Another way that the loss of biodiversity threatens health is through the loss of medicinal plants. Many people depend on plant medicines as their first alternative for treating illness, whether by choice or by lack of access to health services.

Many effective medicines obtained from plants have been known from ancient times. One that is still used is quinine, a malaria treatment that is obtained from cinchona bark, the bark of a South American tree. Quinine gained renewed importance after malaria parasites developed resistance to newer drugs. Now even quinine has become less effective because of resistance in Thailand and neighboring areas, forcing a shift to yet another drug obtained from an ancient Chinese plant remedy.

Morphine for pain relief originates from the opium poppy. Digitalis for the heart is a constituent of the foxglove, a northern European flowering plant. Twentieth-century plant medicine discoveries include most famously the Madagascar rosy periwinkle, source of the anti-cancer alkaloids, vincristine and vinblastine. Steroidal hormones are found in yam tubers. A muscle relaxant is derived from curare, the Amazonian arrow poison of plant origin mentioned earlier in chapter 5.

Animal sources are less common than plants, but people treated for high blood pressure have reason to be grateful to a snake. The model for producing the anti-hypertensive ACE inhibitors was the venom of a Brazilian viper (Balick, Elisabetsky, and Laird 1996).

Many of the important medicinal plants come from tropical forest plants; hence part of the threat of logging these forests is that the source of future pharmaceuticals may be lost. The significance of tropical forests as a source of medicines is not merely coincidental but is related to ecology. The tropics contain a disproportionate percentage of the total number of species. In addition, insects and microorganisms thrive year-round, without a seasonal check due to cold weather. Tropical plants have evolved a host of biochemical defenses against these pests. Many of these substances show chemical activity that is useful to modern medicine and agriculture.

Multinational drug companies do "drug prospecting," deriving clues from the medicinal plants traditionally used by forest peoples. This is a shortcut to choose the plants to screen first. The companies analyze these for active compounds, test them, and then develop synthetic derivatives from them so that they do not need to rely on importing supplies gathered from the forest. Of the top 150 brand-name drugs prescribed in the United States, over half were originally derived from plant and animal sources (Grifo et al. 1997). Increasingly, there is also a profitable market for herbals that have not undergone this full process of testing and synthesis to qualify as drugs but are sold directly as dietary supplements in health stores. Each country has its own standards for herbal medicines. German medical doctors, unlike American doctors, often prescribe herbals such as gingko biloba extract, St. John's wort, and Echinacea.

Research to find new plant medicines is actively carried on by drug companies, universities, botanical gardens, and government laboratories such as those at the National Cancer Institute. Anthropologists have little role in the laboratory side of this process. Their research role is in the field—documenting the context of plant use and studying the cultural classification and explanation of illness. They observe and interview healers, shamans, midwives, and their patients. This kind of study is called ethnomedicine and is part of medical anthropology. If the focus of the study is more on the plants than on the healers, it is referred to as ethnobotany. Ethnobotanists may have their primary training and disciplinary affiliation as either anthropologists or as botanists. In either case, they need to know something of both disciplines.

In the 1990s anthropologists took on a stronger role as advocates for the right of indigenous people to share in the profits to be gained from future drug discoveries. This right applies whether the drug discoveries stem from traditional healers' knowledge or simply from the plants collected from their lands. Initially this concern was described as "intellectual property rights." In common with things such as traditional music, artistic designs, and folktales, it seemed only fair that the plant medicines of indigenous people should be protected by copyright or patent so that they could benefit financially from their knowledge.

It soon became obvious that the concept of intellectual property rights was too focused on economics and litigation to be really helpful to indigenous communities. They were unlikely to have the resources to get a patent or a copyright. It was also difficult to do so because the knowledge was often communally held. The laws were far more likely to work to the advantage of large corporations. Subsequently, anthropologists began to advocate the broader concept of "traditional resource rights," intended to protect biological diversity as well as the cultural knowledge that relates to it (Posey and Dutfield 1996).

EMERGING DISEASES

The loss of medicinal plant species is only one facet of the rapid environmental change that humans are causing. The same forces for change are exposing human populations to new infectious diseases. Pushed by population growth, agriculture and settlements move into new areas. In the process of expansion, predator populations are likely to be reduced through habitat loss and the use of pesticides. These predators might otherwise have kept rodents and insects in check. The expanding populations of rodents and insects may cause crops to be lost or serve as vectors for the spread of infectious disease.

As human populations enter new, marginal areas, they may come into contact with vectors and reservoirs of new infectious diseases for which they have no immunity. Medical anthropologist Mark Nichter (1987) illustrates this series of events with kyasanur forest disease. This disease, caused by a virus spread by ticks, had long been present at low levels in India, but an epidemic broke out after extensive clearing of the forest was done for an economic development project. The cutting of the forest brought people in closer contact with ticks, the insect vector that carries the virus.

Historically, the famous example of altered relationships between people and an urban environment is the Black Death, the epidemic of bubonic plague that swept England and the rest of Europe in the fourteenth century. Fleas are the vectors that transmit the bacteria that cause plague. Normally plague is a disease of rodents, but in early cities large numbers of poor people came in close contact with black rats, attracted by grain stored in their poorly constructed houses.

The close association of people and rats repeats itself in modern American inner cities as the breakdown of city services reduces garbage collection and rat control. Medical tests indicate that that many city residents have antibodies that show a previous exposure to certain viruses for which city rats are the reservoir. These include a hantavirus that may cause kidney damage that eventually shows up as high blood pressure.

Tropical forests and urban jungles may be the first environments that come to mind when we think of disease ecology, but environmental change can lead to outbreaks of disease in arid environments as well. In 1993 a lethal variant of hantavirus was seen in an outbreak in New Mexico. Many of the victims were otherwise healthy young adults tackling spring cleaning in the garage or shed or holding outdoor jobs. Field archaeologists and geologists were two of the occupations especially at risk for exposure. The sick people turned up at emergency rooms with acute flu-like respiratory disease. The animal population that serves as a reservoir for hantavirus in the American Southwest is the deer mouse. In the fall of 1992 there was a population boom among deer mice. The rodent population boom followed a year when heavy rains increased their food supply of pine nuts and grasshoppers.

After the 1993–94 outbreak, hantavirus quieted down for a few dry years until similar conditions led to a renewed outbreak in 1998. These extreme fluctuations in weather may become more common as the trend to global warming continues. As greenhouse gases continue to increase in the atmosphere, particularly the carbon dioxide produced by industrial and automobile emissions, global warming will alter ocean currents, and weather patterns will become even more unstable.

Future global climate change will alter marine ecosystems, that is, the plants and animals living in the oceans. This is bound to affect human health and has already happened in the major cholera epidemics of the 1990s that got started in the warm coastal waters offshore from Peru, India, and Bangladesh. Cholera is a bacterial infection involving the small intestine. It produces severe diarrhea that rapidly leads to dehydration and death unless fluids and electrolytes are replenished. People get cholera by ingesting food or water that has been contaminated with bacteria from the excrement of cholera patients.

Maj-Lis Follér, a Swedish human ecologist, studied cholera among the Shipibo-Conibo Indians of Peru (Follér and Garrett 1996). The 1991 epidemic that started in the coastal cities of Peru spread quickly to the Andes and beyond into the Amazonian lowlands. By April 1991 it had reached the Shipibo-Conibo villages. One of the places where the Shipibo-Conibo were exposed was on the crowded and highly unsanitary ferries, called *collectivos*, the form of public transport that takes them from their villages to the city of Pucallpa. They travel to the city to sell their handicrafts and small cash crops of peanuts and rice and to buy supplies ranging from fuel for their boats to soap and clothing.

Despite their involvement in the cash economy of Peru, the Shipibo-Conibo still rely on traditional healers and herbal medicines, so it was not a foregone conclusion that they would accept Western medicine and its explanations and treatments for cholera. The severity of the diarrhea and the other symptoms of cholera helped them decide that this disease was something alien that called for new explanations and nontraditional treatments. A blitz of health education information by radio and other means from the national health authorities was able to achieve the acceptance of oral rehydration with boiled water, sugar, and salt, to which they added infusions of various medicinal plants. About 250 Shipibo-Conibo out of a total population of 30,000 died in the cholera epidemic. Thanks to good treatment, the death rate was much lower than in cholera epidemics in other places (Follér and Garrett 1996).

The bacteria that cause cholera are of a kind called vibrios, which are curved or spiral-shaped microorganisms. Cholera vibrios thrive among aquatic plants, including seaweed and algae. Warmer sea temperatures increase the growth of algae, thereby increasing the threat of cholera as well.

These coastal algal blooms may be one of the first warning signs of global warming. It is not warmer water alone that encourages the overgrowth of algae, but the combined effect of other changes in the marine ecosystems. The runoff from more intensive human use of the land feeds the algae through soil erosion, fertilizers, sewage, and industrial wastes. Mangroves and wetlands normally filter out some of the excess nitrates and phosphates, but the mangroves and wetlands are being lost to development of resorts and industries along the coast.

Damaged marine ecosystems are simplified ecosystems. They lack the biodiversity that maintains stability. These unstable conditions favor small organisms that multiply rapidly, many of which are toxic or pathogenic. In addition to cholera, another example of an organism causing disease in a marine ecosystem is Pfiesteria, one of the dinoflagellates. Dinoflagellates are shape-shifting organisms that have chlorophyll-like green plants but also have a long tail that lets them swim like animals. Pfiesteria were first identified only a few years before they multiplied enough to kill a billion fish in the coastal waters off North Carolina in 1991. In 1997 Pfiesteria spread north to Chesapeake Bay. The organism produces a toxin that makes fish lethargic and causes bleeding sores in their skin. People do not seem to get sick by eating the diseased fish, but fishermen and lab workers were affected by their exposure to the toxin. The outbreak may be traced to increased pollution of the Atlantic coastal waters by runoff from hog farms in North Carolina and chicken farms along Chesapeake Bay. In contrast to industrial pollution and municipal sewage, controls on agricultural pollution had hardly been considered until the Pfiesteria outbreak.

Pfiesteria research is a good example of the need for multidisciplinary approaches to problems of environmental health. Many of the researchers working on this problem have their primary training in biology, ecology, or epidemiology. Anthropologists have been involved, too, looking at the human aspects of the problem such as the ways that people come in contact with fish and water, the beliefs that people in the communities affected have about the outbreak, and the effects of changing regulations on farmers and fishermen.

Much of the topic of this chapter could well be considered to belong to medical anthropology rather than environmental anthropology. There is a necessary overlap between these two subdisciplines. Many of the health problems that people experience need to be understood by considering the environmental hazards to which they are exposed (McElroy and Townsend 1996). Medical anthropology studies healers, many of whom make use of plant medicines to treat illness and enhance health. Anthropologists who study food and nutrition also work on the interface of environment and health.

The growth of both environmental anthropology and medical anthropology in recent years is related to the development of applied anthropology, the topic of the next chapter. As anthropologists found employment that made practical use of anthropology outside of academic departments, health and the environment were two areas where they were able to contribute their expertise.

Chapter Eleven

It Isn't Easy Being Green

Kermit the Frog, Jim Henson's famous Muppet, used to sing a mournful song that began, "It's not that easy being green." Anthropologists, too, have found a challenge in being "green," that is, to balance personal and political involvement in environmental issues, scholarship that provides understanding of those issues, and the requirements of their employer or sponsoring agency. It is not unusual for an anthropologist to be pulled in contradictory directions. We can easily imagine a few fictionalized examples drawn from real situations:

> A researcher has spent some years studying a hunting-and-gathering people in Southeast Asia who have been resisting the clear-cut logging of their lands. The case has attracted worldwide media attention. The anthropologist is disturbed by the way the indigenous people are portrayed in the media as primitive, helpless, and animal-like, because anyone who knows them well would see them as politically sophisticated human beings with aspirations for education, modern health care, and economic development on their own terms. The problem is that the romantic media image of them as victims succeeds in building the support they need more than the corrected picture would (Brosius 1999).

> Another anthropologist works with a well-funded epidemiological team on the problem of Pfiesteria (discussed in chapter 10). Their research has shown that the frequency and severity of its effects on human health in the coastal areas are quite low, making it a rather minor public health problem at the present time. Though other researchers disagree, he regards it as relatively insignificant compared to the big problem of occupational injuries to low-paid workers in the food-processing industries of the area. A journalist goes after the anthropologist for anecdotes about Pfiesteria, which has captured the imagination of a public whose appetite is whetted by movies and novels about outbreaks of mysterious diseases. The newspaper's readers don't really want to hear about routine health problems of poor people (Griffith 1999).

Ecological anthropologists who went to do fieldwork in the past most often studied the adaptation of human communities to their biotic and physical environment. Ecological anthropologists who go to do fieldwork now are just as likely to go to the offices of an environmental organization to observe and interview staff and volunteers. Even in a tropical village setting, rather than weighing sweet potatoes and measuring gardens, they are as likely to attend a meeting of villagers deciding how to manage their forest resources in response to an offer to sell their logs to a timber company.

Increasingly ecological anthropologists seek employment as applied anthropologists, working for institutions and organizations concerned with the environment. Persons with degrees in anthropology are hired to do full-time work that does not have the job title "anthropologist," but they use their social science skills on the job as administrators or project evaluators. Sometimes anthropologists working full time in universities are sought as temporary consultants to advise on a project in their area of expertise.

A characteristic method that applied anthropologists use to gain understanding of a practical problem is to begin at the local level and progressively explore linkages to regional, national, and global levels. One of the anthropologists who has best bridged the world of academic and applied anthropology is Andrew Vayda. Vayda had been Roy Rappaport's academic advisor for his work in New Guinea, but their later development of ecological anthropology diverged. Some of Vayda's theoretical ideas grew out of his very practical attempts over the years to understand and explain deforestation in Indonesia.

THE INDONESIAN FOREST FIRES OF 1997–98

In June and July, 1998, Vayda traveled to Indonesia as a consultant to the World Wide Fund for Nature (WWF)—Indonesia. This is a local branch of WWF, which is a big international environmental organization known by its panda logo. Working alongside an Indonesian colleague, Ahmad Sahur, Vayda tried to explain the causes of the devastating forest fires that followed the drought of 1997–98 (Vayda 1999). (This was the same widespread drought that temporarily shut down the Ok Tedi Mine in Papua New Guinea, as discussed in chapter 7.) The burning of dry vegetation created a smoky haze that closed airports and caused respiratory problems all over Southeast Asia. Much of the smoke haze was caused by fires that were deliberately set to clear land for large agricultural projects that included a huge government-supported rice project as well as plantations of oil palms and other tree crops. This land would have been cleared even if the drought had not occurred. The WWF, Vayda, and Sahur were interested in the cause of the other fires, the forest fires that would not have happened had it not been for the extraordinarily dry conditions.

Government bureaucrats prefer to blame local farmers clearing land for cultivation as the scapegoats in setting fires that get out of control. Vayda could find no evidence that this was true. In areas under shifting cultivation, farmers were experienced in keeping fires for land clearing under control. Any well-established community must have sanctions (rules and fines or other punishments) to prevent people from letting fires get out of control and damaging the fields and houses belonging to their neighbors.

Field research on the cause of fires led Vayda to look more closely at the illegal practice of cutting ironwood trees (*ulin*) for their fire-resistant, strong, and durable wood that is highly valued for building houses. The ulin cutters used fire in several ways. They set fires to clear the underbrush to make it easier to cut and move the timber. They set small fires for their own comfort—to cook food, smoke cigarettes, and repel mosquitoes. Any of these fires could get out of control. An underlying cause for the forest fires was that previous logging had left degraded forests more susceptible to burning than areas that had never been logged.

Trucking was done by night to evade the police. Truck drivers moving the cut timber along forest roads at night sometimes had to stop for repairs or to change a flat tire. They lighted fires to see by and to scare away mosquitoes or ghosts. Like the fires set by the wood cutters, any of these small fires might get out of control under the dry conditions.

The illegal ironwood trade was a more attractive activity than it might otherwise have been, with all its risks, because Indonesia was in an economic crisis with low wages and high unemployment. Because the trade was illegal, the researchers needed to use indirect methods to estimate its extent. They estimated the number of boatloads of timber that left the port nearest the forest. Then they calculated how many trees would need to be cut down to fill these boats and how many truckloads needed to ship the logs to the port.

The research on forest fires was intended to form a solid basis for policy both on the part of the environmental organization (WWF) and the Indonesian government. Obviously if either organization based its policies on the erroneous assumption that most fires were caused by village farmers clearing their fields, these policies would miss the mark rather badly. In the next case we will see how a campaign carried out by Greenpeace and other environmental organizations *did* have serious ramifications for indigenous people.

ANIMAL RIGHTS IN THE NORTH

Following intense protests and boycotts, the animal rights movement successfully achieved a ban on the import of sealskin into Europe—the world's most important fashion market. The keystone of the campaign was a series of TV and magazine ads showing the slaughter of baby harp seals in Newfoundland. The campaign did not distinguish this commercial southern hunt from the aboriginal seal hunt of the far north. Nor did it make any distinction among the various species of seals. The decision to ban sealskin imports did not rest on scientific study of the environmental issues that might allow a sustainable harvest of seals. The controversy was about animal rights versus human rights, though this distinction is often blurred. Only tangentially was it about the environment. In any case, the ban had a serious economic impact on the Canadian Inuit, for whom the hunting of animals, particularly the ringed seal, continues to have importance. In order to understand the northern seal controversy, an anthropologist needs to approach it from two angles: looking at the place of seal hunting in Inuit society and looking at the animal rights movement as a product of its cultural context in Europe and America (Wenzel 1991).

The "traditional" nature of Inuit hunting is not a matter of technology, for this has changed in the past few decades to incorporate outboard motors, snowmobiles, and rifles in place of sealskin boats, dog teams, and harpoons. Nor is it a matter of traditional economics, for in addition to continuing to eat seal meat, the Inuit now require cash for ammunition, fuel, spare parts, and many other things. The traditional nature of the hunt lies in culture history, social relationships, and the relationship between people and animals. For the Inuit, encounters with animals are governed by moral relationships like those with humans. All relationships with sentient beings are relationships of respect and reciprocity. The social aspects of hunting are particularly important—who hunts with whom and how meat is shared. Socially and culturally, the Inuit have not ceased to be Inuit simply by being incorporated into the market economy.

When anthropologists seek to be advocates for the rights of indigenous people, as Wenzel did for the Inuit (or as Kirsch did for the Yonggom confronted with damage from mine wastes), it is not sufficient that they understand the indigenous people and their environment. The anthropologists also run up against their own culture's contradictory views of indigenous people. The Noble Savage view of the Inuit is that they are isolated people with superb survival strategies for meeting the challenge of an extreme environment and therefore deserving of paternalistic protection. If they are not that, then they must be fully modern and "just like us." Part of what anthropologists need to do is to try to correct this double vision, to create an image of Inuit as "a traditional people living in the modern world," as Wenzel (1991, p. 84) puts it, or as bicultural people who fully and flexibly move between two worlds.

When the Makah Indians of the Olympic Peninsula in the state of Washington decided to hunt a gray whale in 1999 for the first time in 70 years, the uproar in the media raised many of the same issues as the harp seal controversy. The 1855 Treaty of Neah Bay guaranteed the Makah the right to hunt whales and seals at their usual grounds, but they gave up the hunt voluntarily for many years. At one time whales were threatened with extinction by commercial whaling, but by 1999 the population of gray whales was healthy and growing. The Makah decided that they could hunt a whale, not because they especially needed it for food but as part of the revival of their culture and its rituals. Environmental groups did not oppose the Makah hunt, but animal rights activists opposed it very vocally, as did other individuals. Some of them smeared anti-Indian slogans on rally banners, Internet postings, and letters to the editor of newspapers.

ENVIRONMENTAL MOVEMENTS

Both the animal rights movement and environmentalism, whether expressed as individual concerns or as organized social movements, are part of the culture in which they are found. Although there are international environmental organizations, still environmentalism takes different forms and has different histories in the United States and Germany, for example, even though both are wealthy industrial countries. Environmental movements among rural people in developing countries, such as the Chipko movement to protect forests in India, take still different shapes. The political alliances that form between international environmental NGOs and indigenous minorities are often complex and fraught with mutual misunderstandings. One of the most useful things that anthropology can do is to study environmentalism itself and its expression in different cultures (Milton 1996).

An environmental anthropologist is not necessarily also an environmental activist. An activist is motivated to work to protect the environment through political engagement and changes in personal consumption. Probably most environmental anthropologists would not choose to specialize in that area of study if we were not environmentally concerned, but it is certainly possible to do so, just as researchers may study religion or marriage without themselves being believers or married people.

THE QUESTION OF THE COMMONS

Many of the resources we have considered in this chapter or indeed throughout this text—Indonesian forests, hunting rights to sea mammals, water for irrigation—were not privately owned. Instead they were held in common by a group of people, small or large. Anthropologists have done a good deal of research over the years that looked at questions of the ownership of property. But it was an ecologist, not an anthropologist, Garrett Hardin, who proposed the controversial and influential theory of "the tragedy of the commons" in 1968. Hardin's view was a pessimistic one based on a simple picture of an old English commons, a shared pasture. Hardin asserted that each herdsman will selfishly keep on adding more cows to his herd because he gets the full benefits of each cow and calf while the whole community shares the costs of overgrazing, the damage to the pasture.

Economists further developed and applied Hardin's model, arguing that only private property could protect resources. In this bleak view, humans are doomed to overpopulate and degrade the environment because individuals will choose their private advantage over the common good. Anthropologists criticized the individualistic bias of the commons model, showing that property rights around the world are much more complex and embedded in historical and social contexts (McCay and Acheson 1987). Traditionally, people succeeded in managing common-pool resources in sustainable ways for thousands of years. "Common property" does not mean that anybody and everybody has open access; instead societies have systems of rights, duties, and obligations that protect resources held in common. Furthermore, owners of private property do not always behave in ecologically responsible ways either, often discounting future use in favor of present gain or simply making mistakes in managing their resources.

The environmental problems that we now face, with a world population of 6 billion, are related to common-pool resources on an international or global scale. These include such things as the fresh water of the Great Lakes on the U.S.–Canadian border, the ozone layer, and the global climate. Anthropologists traditionally have not dealt with such large-scale questions, preferring to work at the local level. In the final chapter we will ask whether things that anthropologists have learned in their studies of small-scale systems have any relevance for developing policies to address global problems.

Chapter Twelve

Consumer Cultures

When we first returned to the United States from fieldwork in Papua New Guinea, we experienced so-called *reverse culture shock*. At first I kept asking, "Why is everyone all dressed up in their good clothes? Is something special happening?" My husband, Bill, was overwhelmed by the size of the supermarket and the huge number of choices there. After we made these initial adjustments to a home culture where we no longer seemed entirely at home, deeper questions remained. One of these is, "Can the Saniyo teach us anything about how to live sustainably?"

Some ecological anthropologists say yes. They are confident that human systems of knowledge and action that are ecologically sensitive can evolve in the future. Studies of very different ways of life such as that of the Culina of Peru or the Saniyo-Hiyowe of Papua New Guinea contribute to our understanding of the evolutionary or adaptive processes that lead to sustainability and will contribute to this goal.

Other anthropologists say no, forest peoples have little to teach us. Modern life is different in principle, and there is no going back. Dominated by market forces, we now see nature as an impersonal source of resources and services. We no longer see nature as a personal being or beings with whom we might be in relationship. Nature is "it," not "thou."

My own sense is that neither of these is quite true: I am not optimistic about our ability, as a species, to learn to live sustainably before irreversible damage is done to the biosphere—damage of a magnitude sufficient to lead to our own extinction as a species. Even so, pessimism does not get us off the hook as individuals, families, and communities to do our best to change this. Such change may require us to change our own lifestyle and to advocate for change in policies in our local community and our nation. Along these lines, it seems to me that the most important thing the Saniyo have to teach us is to recognize the *abnormality* of our high-consumption way of life. By experiencing other cultures, even in books or film, we may be able to keep alive a sense of reverse culture shock like that Bill and I experienced in returning home. In other words, we may become more aware of the size of our *ecological footprint*.

ECOLOGICAL FOOTPRINT ANALYSIS

The concept of an ecological footprint was not invented by anthropologists, but it is consistent with the way that ecological anthropologists compare the carrying capacity of lands under different systems of food production. In addition to land used for food production, the ecological footprint includes land needed for forest products and the disposal of wastes, in order to estimate the total load on the ecosphere (Wackernagel and Rees 1996).

The ecological footprint of the average North American living in the United States or Canada measures four to five hectares, more than three city blocks. The average footprint of a person in India is less than a tenth of that (.38 hectares). Even so, the total for India is more than the total area of India's productive land. How can this be? It means that India's economy is not sustainable. India is a net importer of food and is depleting its natural capital by the cutting its forests. Most industrial nations are also consuming more than their ecologically productive area is able to produce. They meet this deficit by importing resources from poor countries and depleting their natural capital. In other words, today's industrial economies are not sustainable.

Humans have taken over more than half of the world's resources for their own use, leading some ecologists to speak of human domination of Earth's ecosystems. Humans use more than half of the world's accessible surface fresh water, much of it in agriculture. When surface water is unavailable, they draw on ground water. Some of this is so-called fossil water, meaning that it is nonrenewable—for example, three-quarters of the water currently used in Saudi Arabia is fossil water. Two-thirds of the marine fisheries are overexploited or at the limit of exploitation. Nearly half of the land has been transformed by agriculture, grazing, or the harvest of forests for fuel or timber. At least 40 percent of the annual net growth of plants is used by humans, leaving less than 60 percent for all other species (Vitousek et al. 1997).

The intensity of human use of the planet is due to the increasing scale of operations of the economy as well as to population growth. Another doubling of the world population, which is likely to occur in our lifetime, would exceed many of the resources available to support human life, requiring changes in the present pattern of greedy consumption.

EUROPEAN AND AMERICAN CONSUMERS

A cultural anthropologist does not have to go as far as New Guinea to observe differences in consumption and lifestyle. Rita Erickson did her fieldwork in small towns in Minnesota and Sweden in the early 1980s and again in the 1990s. The towns were similar, as were their daily life routines. She was interested in looking at residential energy use, because nationally the Swedes used only 70 to 80 percent per person of what Americans used, enough to make a difference in their contribution to global environmental problems (Erickson 1997).

Surprisingly she found the Swedes actually used more heating fuel, keeping their thermostats up at night and ventilating the house by opening windows and doors. The Americans made more use of appliances; watched more TV; and owned more air conditioners, humidifiers, and clothes driers. Energy conservation in both communities had been encouraged by high fuel prices in the late 1970s. As prices stabilized, energy use crept up in both communities.

In other environmental behaviors, the two communities differed, too. Minnesota is a leading state in recycling in the United States, but the residents of the Swedish town were far more involved in recycling glass and paper. The Swedish families were also more likely to compost yard or kitchen waste and less likely to use pesticides, which are strictly regulated in Sweden.

What the Americans and Swedes share is that both belong to industrial consumer societies. Although they may express disgust at materialism and consumerism, they carry on acquiring more. Status symbols in the 1990s in both communities include vehicles for adults, label clothing for teens, and video games for kids. Americans favored big homes in new developments while Swedes emphasized impressive yards and high-quality home furnishings. Consumer spending on goods and services is about two-thirds of gross domestic product in both the United States and Sweden (Erickson 1997, p. 113).

These consumer goods can be seen as "embodied energy." They represent the use of energy in manufacturing, distribution, and disposal. All of these processes burden the environment. It has been estimated that it would take at least three planets for everyone now living on earth to live like this (Wackernagel and Rees 1996, p. 15). At various times in the past, environmentalists have suggested that the limits of human consumption would be reached when nonrenewable resources such as petroleum and minerals were used up. Now it seems more likely that the limiting factor will be fresh water as well as places to dispose of waste. As technology is developed to cope with one shortage, another takes its place. Each of the "solutions" brings in its wake a new set of costs and problems, demanding further human adaptability, increasing inequities, and diminishing the quality of life.

This text has discussed environmental anthropologists at work in various settings, describing relationships between humans and their environments in modern cities, farming villages, and foraging bands, past and present. Some of the anthropologists have been academics working toward better methods and theories; others have been applied anthropologists trying to solve practical local problems.

More recently anthropologists have begun to address *global* environmental problems such as the climate change that is coming about from the burning of greenhouse gases. There is a precedent for this, in that archaeologists point out that people have been modifying landscapes through the use of fire in hunting for thousands of years. Humans have now affected all of the nonhuman biosphere, so that there is nowhere to go, even in the interior mountains of New Guinea, to escape the effects of technologies such as pesticides. Anthropologists themselves have enlarged their frame of reference to encompass global processes, and nonanthropologists have begun to realize that anthropologists hold some keys to understanding the way that humans are able to adapt to a shrinking world.

Glossary

Adaptation. A process of change or adjustment that is beneficial for a population, making individual organisms more suited to the stresses of their environment.

Agriculture. Farming of an intensive type using irrigation and plows drawn by animals or tractors.

Applied anthropology. The use of anthropology to solve practical problems.

Archaeology. The subfield of anthropology that studies the material remains of people of the past.

Biodiversity. The variety of life, including the full range of variation in species, ecosystems, populations, and genes.

Biological anthropology. The subfield of anthropology that studies the physical (biological) aspects of the human species.

Carrying capacity. The size of population that can be supported by an area of land.

Caste. A hereditary social group associated with a particular occupation in a hierarchy of such groups.

Circumscription. The fact of being enclosed or surrounded, often by a geographical barrier.

Cognition, cognitive. Having to do with perceiving and thinking.

Cultural anthropology. The subfield of anthropology that studies the ways of life of contemporary people.

Cultural ecology. A theoretical perspective that emphasizes the importance of the natural environment in shaping core features of culture, including technology and economics.

Cultural relativism. The principle that cultures are to be evaluated in terms of their own values and not those of another culture.

Culture. A way of life; all that people learn to do, say, make, and think as members of society.

Ecological anthropology. The study of relationships between a population of humans and their biophysical environment.

Ecological footprint. The area of land that would be required to support a defined human population at its current material standard of living indefinitely; a measure of the load placed on the environment by using resources and disposing of wastes.

Ecosystem. A system formed by interactions within a community of different species of organisms, including humans, and its biophysical environment, characterized by flows of information, energy, and matter.

Environmental anthropology. The use of anthropology's methods and theories to contribute to the understanding of local or global environmental problems.

Equilibrium. Balance.

Ethnoecology. The study of the knowledge and beliefs about nature that are held in a particular culture, a broad cover term that includes such subdisciplines as ethnobotany, ethnozoology, and ethnobiology.

Feedback. A response that is used to alter a later response.

Foraging. Subsistence from the wild through obtaining food by hunting and gathering.

Hacienda. (Spanish) A large ranch or farm owned by a member of the elite class.

Hantavirus. A recently discovered genus of viruses that are spread to humans from rodent urine, causing severe disease (see also *Virus*).

Hazard. Something that causes danger or risk, especially of injury or death.

Homeostasis. Maintaining the same internal conditions despite changes in external conditions.

Horticulture. Farming of a less intensive kind using hand tools. A long fallow period restores the land without need for irrigation or fertilizers.

Kilocalorie. A measure of the energy-producing value in food when oxidized by the body, equivalent to 1 Calorie (also called a "large calorie") or 1,000 calories.

Linguistics. The science of language. As a subfield of anthropology, anthropological linguistics is primarily concerned with systematically describing a language as a part of the culture in which it occurs.

Niche (ecological). A species' distinctive way of living, using resources, and relating to competitors, its specialized role in the environment.

Nongovernment organizations. Voluntary organizations such as charitable and environmental organizations.

Optimal foraging model. A theory that predicts that humans and other predators will hunt the species that provide the best short-term return for their effort, regardless of cultural preferences or long-term effects.

Papua New Guinea. A country in the South Pacific that became independent from Australia in 1975. It occupies the eastern half of the island of New Guinea.

Political ecology. An orientation to research in environmental studies that espouses the viewpoint that relationships between humans and their environment cannot be understood without considering inequalities of power and wealth, especially those produced by the global economy.

Population. An interacting local group of individuals within which most mating takes place.

Reservoir. A population of animals that maintains a parasite or other pathogen from which humans may become infected.

Risk. The likelihood of some event, such as acquiring a disease or experiencing an accident.

Savanna. A tropical grassland environment.

Semantic domain. A set of terms in a language within a particular area of knowledge, such as a set of kinship terms or plant names.

Sustainability. The degree to which a given practice or material standard of living can continue without using up the ability to do so in the future.

System. A set of objects and their relationships.

Systems theory. A set of analogies drawn from computers and electronics and applied to biology, ecology, and other fields to understand communication and the flow of energy and information. Such cybernetic models differ from simple mechanical models by incorporating *feedback*.

Vector. An insect that is a carrier of a disease organism.

Virus. A microorganism consisting of nucleic acid in a protein capsule that can grow and replicate itself only within other cells.

References

Almendares, J., et al. 1993. "Critical Regions, a Profile of Honduras." *Lancet* 342(8884): 1400–1402.

Alvard, Michael. 1995. "Intraspecific Prey Choice by Amazonian Hunters." *Current Anthropology* 36(5): 789–818.

Anderson, Eugene N. 1996. *Ecologies of the Heart: Emotion, Belief, and the Environment*. New York: Oxford University Press.

Balée, William. 1999. "Mode of Production and Ethnobotanical Vocabulary: A Controlled Comparison of Guajà and Ka'apor." In *Ethnoecology: Knowledge, Resources, and Rights*, ed. T. L. Gragson and B. G. Blount, pp. 24–40. Athens: University of Georgia Press.

Balick, Michael J., Elaine Elisabetsky, and Sarah A. Laird. 1996. *Medicinal Resources of the Tropical Forest: Biodiversity and Its Importance to Human Health*. New York: Columbia University Press.

Barth, Fredrik. 1958. "Ecological Relationships of Ethnic Groups in Swat, North Pakistan." *American Anthropologist* 58: 107–189.

Bennett, John William. 1975. *The Ecological Transition: Cultural Anthropology and Human Adaptation*. New York: Pergamon Press.

Bodley, John H. 1999. *Victims of Progress.* Mountain View, CA: Mayfield.

Brosius, J. Peter. 1999. "Analyses and Interventions: Anthropological Engagements with Environmentalism." *Current Anthropology* 40(3): 277–309.

Button, Gregory V. 1995. "What You Don't Know Can't Hurt You: The Right to Know and the Shetland Island Oil Spill." *Human Ecology* 23(2): 241–258.

Carneiro, Robert L. 1970. "A Theory of the Origin of the State." *Science* 169: 733–738.

Carson, Rachel. 1962. *Silent Spring*. Boston: Houghton Mifflin.

Cole, John W., and Eric R. Wolf. 1974. *The Hidden Frontier: Ecology and Ethnicity in an Alpine Valley*. New York: Academic Press.

Conklin, Harold C. 1954 "An Ethnoecological Approach to Shifting Agriculture." New York Academy of Sciences, *Transactions Series* 2(17): 133–142.

Crumley, Carole L. 1993. *Historical Ecology: Cultural Knowledge and Changing Landscapes*. Santa Fe, NM: School of American Research Press.

Descola, Philippe. 1994. *In the Society of Nature: A Native Ecology in Amazonia*. Cambridge, U.K.; New York; Paris: Cambridge University Press; Editions de la Maison des sciences de l'homme.

———. 1996. *The Spears of Twilight: Life and Death in the Amazon Jungle*. New York: New Press.

Dove, Michael R. 1993. "Uncertainty, Humility, and Adaptation in the Tropical Forest: The Agricultural Augury of the Kantu'." *Ethnology* 32(2): 145–168.

Durham, William H. 1995. "Political Ecology and Environmental Destruction in Latin America." In *The Social Causes of Environmental Destruction in Latin America*, ed. M. Painter and W. H. Durham, pp. 249–264. Ann Arbor: University of Michigan Press.

Early, John D., and Thomas N. Headland. 1998. *Population Dynamics of a Philippine Rain Forest People: The San Ildefonso Agta*. Gainesville: University Press of Florida.

Edgerton, Robert B. 1992. *Sick Societies: Challenging the Myth of Primitive Harmony*. New York: Free Press.

Ehrlich, Celia. 1999. "The Ethnobotany of *Cordyline fruticosa (L.) A Chev*: The 'Hawaiian Ti Plant.'" Ph.D. diss., State University of New York at Buffalo.

Ehrlich, Paul R., and Anne H. Ehrlich. 1990. *The Population Explosion*. New York: Simon and Schuster.

Ellen, Roy F. 1982. *Environment, Subsistence, and System: The Ecology of Small-scale Social Formations*. New York: Cambridge University Press 1982.

Erickson, Rita J. 1997. *"Paper or Plastic?": Energy, Environment, and Consumerism in Sweden and America*. Westport, CT: Praeger.

Escobar, Arturo. 1998. "Constructing Nature: Elements for a Poststructural Political Ecology." In *Liberation Ecologies: Environment, Development, Social Movements*, ed. R. Peet and M. Watts, pp. 46–68. London: Routledge.

Fedick, Scott L. 1996 *The Managed Mosaic: Ancient Maya Agriculture and Resource Use*. Salt Lake City: University of Utah Press.

Follér, Maj-Lis, and Martha J. Garrett. 1996. "Modernization, Health and Local Knowledge: The Case of the Cholera Epidemic among the Shipibo-Conibo in Eastern Peru." In *Human Ecology and Health: Adaptation to a Changing World*, ed. M.-L. Follér and L. O. Hansson, pp. 135–166. Göteborg, Sweden: Göteborg University, Department of Interdisciplinary Studies of the Human Condition.

Franke, Richard W., and Barbara H. Chasin. 1994. *Kerala: Radical Reform as Development in an Indian State*. Oakland, CA: Institute for Food and Development Policy.

Fratkin, Elliot M., Cultural Survival, Inc., and Smith College. 1998. *Ariaal Pastoralists of Kenya: Surviving Drought and Development in Africa's Arid Lands*. Boston: Allyn and Bacon.

Geertz, Clifford. 1963. *Agricultural Involution*. Berkeley: Published for the Association of Asian Studies by University of California Press.

Griffith, David. 1999. "Exaggerating Environmental Health Risk: The Case of the Toxic Dinoflagellate Pfiesteria." *Human Organization* 58(2): 119–128.

Grifo, Francesca, et al. 1997. *Biodiversity and Human Health*. Washington, DC: Island Press.

Hardesty, Donald L. 1977. *Ecological Anthropology*. New York: Wiley.

Hill, Kim, and A. Magdalena Hurtado. 1996. *Aché Life History: The Ecology and Demography of a Foraging People.* New York: Aldine de Gruyter.

Howell, Nancy. 1979. *Demography of the Dobe !Kung*. New York: Academic Press.

Hunn, Eugene. 1989. "Ethnoecology: The Relevance of Cognitive Anthropology for Human Ecology." In *The Relevance of Culture*, ed. M. Freilich, pp. 145–160. New York: Bergin & Garvey.

———. 1999. "Ethnobiology in Court: The Paradoxes of Relativism, Authenticity, and Advocacy." In *Ethnoecology: Knowledge, Resources and Rights,* ed. T. L. Gragson and B. G. Blount, pp. 1–11. Athens: University of Georgia Press.

Hunn, Eugene S., and James Selam. 1990. *Nch'i-wána, "The Big River": Mid-Columbia Indians and Their Land*. Seattle: University of Washington Press.

Hyndman, David. 1994. *Ancestral Rain Forests & the Mountain of Gold: Indigenous Peoples and Mining in New Guinea*. Boulder, CO: Westview Press.

Keen, David. 1994. *The Benefits of Famine: A Political Economy of Famine and Relief in Southwestern Sudan, 1983–1989*. Princeton, NJ: Princeton University Press.

Kirsch, Stuart. 1995. "Social Impact of the Ok Tedi Mine on the Yonggom Villages of the North Fly, 1992." *Research in Melanesia* 19: 23–102.

Krauss, Michael. 1992. "The World's Languages in Crisis." *Language* 68(1): 4–10.

Lansing, John Stephen. 1991. *Priests and Programmers: Technologies of Power in the Engineered Landscape of Bali*. Princeton, NJ: Princeton University Press.

Lansing, J. Stephen, and James N. Kremer. 1993. "Emergent Properties of Balinese Water Temple Networks: Co-adaptation on a Rugged Fitness Landscape." *American Anthropologist* 95(1): 97–114.

Little, Michael A., and George E. B. Morren. 1976. *Ecology, Eenergetics, and Human Variability*. Dubuque, IA: W. C. Brown.

MacCormack, Carol P., and Marilyn Strathern. 1980. *Nature, Culture, and Gender*. New York: Cambridge University Press.

Martin, Paul S., and Richard G. Klein. 1984. *Quaternary Extinctions: A Prehistoric Revolution*. Tucson: University of Arizona Press.

McCay, Bonnie J., and James M. Acheson. 1987. *The Question of the Commons: The Culture and Ecology of Communal Resources*. Tucson: University of Arizona Press.

McElroy, Ann, and Patricia K. Townsend. 1996. *Medical Anthropology in Ecological Perspective*, Third Edition. Boulder, CO: Westview Press.

McGuire, Thomas R. 1997. "The Last Northern Cod." *Journal of Political Ecology* 4: 41–54.

Mealey, George A. 1996. *Grasberg*. New Orleans: Freeport-McMoRan Copper & Gold.

Milton, Kay. 1996. *Environmentalism and Cultural Theory: Exploring the Role of Anthropology in Environmental Discourse*. London; New York: Routledge.

Moran, Emilio F. 1979. *Human Adaptability: An Introduction to Ecological Anthropology*. North Scituate, MA: Duxbury Press.

———. 1993. *Through Amazonian Eyes: The Human Ecology of Amazonian Populations*. Iowa City: University of Iowa Press.

Nazarea, Virginia D. 1998. *Cultural Memory and Biodiversity*. Tucson: University of Arizona Press.

Netting, Robert McC. 1981. *Balancing on an Alp: Ecological Change and Continuity in a Swiss Mountain Community*. Cambridge: Cambridge University Press.

———. 1986. *Cultural Ecology*, 2nd ed. Prospect Heights, IL: Waveland Press.

Nichter, Mark. 1987. "Kyasanur Forest Disease: An Ethnography of a Disease of Development." *Medical Anthropology Quarterly* 1: 406–423.

Odum, Eugene Pleasants. 1953. *Fundamentals of Ecology*. Philadelphia: Saunders.

Posey, Darrell, et al. 1984. "Ethnoecology as Applied Anthropology in Amazonian Development." *Human Organization* 43(2): 95–107.

Posey, Darrell Addison, and Graham Dutfield. 1996. *Beyond Intellectual Property: Toward Traditional Resource Rights for Indigenous Peoples and Local Communities*. Ottawa: International Development Research Centre.

Rappaport, Roy A. 1984. *Pigs for the Ancestors: Ritual in the Ecology of a New Guinea People*, 2nd ed. New Haven, CT: Yale University Press. Reissued Prospect Heights, IL: Waveland Press, 2000.

Rathje, William L., and Cullen Murphy. 1992. *Rubbish!: The Archaeology of Garbage*. New York: HarperCollins.

Roosevelt, Anna. 1989. "Resource Management in Amazonia before the Conquest: Beyond Ethnographic Projection." *Advances in Economic Botany* 7: 30–62.

Ross, Eric Barry. 1978. "Food Taboos, Diet, and Hunting Strategy: The Adaptation to Animals in Amazon Cultural Ecology." *Current Anthropology* 19(1): 1–36.

Ross, Eric B. 1986. "Potatoes, Population, and the Irish Famine: The Political Economy of Demographic Change." In *Culture and Reproduction: An Anthropological Critique of Demographic Transition Theory*, ed. W. P. Handwerker, pp. 196–220. Boulder, CO: Westview Press.

Ruddle, Kenneth, et al. 1978. *Palm Sago: A Tropical Starch from Marginal Lands*. Honolulu: Published for the East-West Center by the University Press of Hawaii.

Sahlins, Marshall David, Elman Rogers Service, and Thomas G. Harding. 1960. *Evolution and culture*. Ann Arbor: University of Michigan Press.

Shaw, Rosalind. 1992. "'Nature,' 'Culture' and Disasters: Floods and Gender in Bangladesh." In *Bush Base: Forest Farm*, ed. E. Croll and D. Parkin, pp. 200–217. London: Routledge.

Smith, Eric Alden. 1991. *Inujjuamiut Foraging Strategies: Evolutionary Ecology of an Arctic Hunting Economy*. New York: A. de Gruyter.

Sponsel, Leslie E. 1997. "The Master Thief: Gold Mining and Mercury Contamination in the Amazon." In *Life and Death Matters: Human Rights and the Environment at the End of the Millennium*, ed. B. R. Johnston, pp. 99–127. Walnut Creek, CA: AltaMira Press.

Steward, Julian Haynes. 1955. *Theory of Culture Change: The Methodology of Multilinear Evolution*. Urbana: University of Illinois Press.

Stonich, Susan C. 1955. *"I Am Destroying the Land!": The Political Ecology of Poverty and Environmental Destruction in Honduras*. Boulder, CO: Westview Press.

Storey, Rebecca. 1992. "Children of Copán: Issues in Paleopathology and Paleodemography." *Ancient Mesoamerica* 3(1): 161–167.

Thomas, R. Brooke. 1976. "Energy Flow at High Altitude." In *Man in the Andes: Multidisciplinary Study of High-altitude Quechua*, ed. P. T. Baker and M. A. Little, pp. 379–404. Stroudsburg, PA: Dowden Hutchinson & Ross.

———. 1997. "Wandering toward the Edge of Adaptability: Adjustments of Andean People to Change." In *Human Adaptability Past, Present, and Future: The First Parkes Foundation Workshop, Oxford, January 1994*, ed. S. J. Ulijaszek and R. Huss-Ashmore, pp. 183–232. New York: Oxford University Press.

Vayda, Andrew P. 1999. *Finding Causes of the 1997–98 Indonesian Forest Fires: Problems and Possibilities*. Jakarta: World Wide Fund for Nature—Indonesia. Website: www.wwf.or.id.

Vitousek, Peter M., et al. 1997. "Human Domination of Earth's Ecosystems." *Science* 277: 494–499.

Wackernagel, Mathis, and William E. Rees. 1996. *Our Ecological Footprint: Reducing Human Impact on the Earth*. Gabriola Island, BC: New Society Publishers.

Webster, David. 1999. "The Archaeology of Copán, Honduras." *Journal of Archaeological Research* 7(1): 1–53.

Webster, David, AnnCorinne Freter, and Nancy Gonlin. 2000. *Copán: The Rise and Fall of an Ancient Maya Kingdom.* Fort Worth, TX: Harcourt.

Wenzel, George W. 1991. *Animal Rights, Human Rights: Ecology, Economy, and Ideology in the Canadian Arctic.* Toronto: University of Toronto Press.

Whittington, Stephen L., and David M. Reed. 1997. *Bones of the Maya: Studies of Ancient Skeletons.* Washington, DC: Smithsonian Institution Press.

Wilson, Edward O.1989. "Threats to Biodiversity." *Scientific American* (September): 108–116.

Wolf, Eric. 1972. "Ownership and Political Ecology." *Anthropological Quarterly* 45: 201–205.

Wright, Lori E., and Francisco Chew. 1999. "Porotic Hyperostosis and Paleoepidemiology: A Forensic Perspective on Anemia among the Ancient Maya." *American Anthropologist* 100(4): 924–939.

Index